IDEAFEST JOURNAL

THE INTERDISCIPLINARY JOURNAL OF CREATIVE WORKS AND RESEARCH FROM HUMBOLDT STATE UNIVERSITY

Humboldt State University Press

Humboldt State University
1 Harpst St.
Arcata CA 95521-8299

digitalcommons.humboldt.edu/ideafest

ISBN-13: 978-1-947112-29-2

CONTENTS

IDEAFEST JOURNAL
VOLUME 3 • SUMMER 2019

ASTRONOMY

BIOLOGICAL SCIENCES

GEOSPATIAL SCIENCES

RELIGIOUS STUDIES

RHETORIC

SOCIOLOGY

CONTENTS

LETTER FROM THE EDITOR

Welcome to the third edition of *ideaFest Journal*! I am thrilled to be involved for the second year as managing editor. It has been a privilege to work with so many talented and driven authors, including undergraduates, graduate students, and faculty. I would like to extend a special thanks to Kyle Morgan, who gave me the freedom and time to make this journal whatever I wanted it to be.

The journal has experienced significant growth over the past year, with more than twice as many publications in this 2019 volume. Not only have we published more articles, but we have had the opportunity to introduce three new fields of interest: religious studies, sociology, and rhetoric. With expansion in the quantity and diversity of submissions, we want to thank the peers and specialists in each respective field that provided peer-reviews. Thanks to these peer-reviewers, we were able to publish high-quality articles and give authors the experience of incorporating professional critiques into their work.

As you, the readers, make your way through this issue, please consider how you might contribute to future editions of ideaFest Journal and the legacy of a thriving university publication. Contributions come in many shapes and forms. Do you have a semester-long project with a write-up? Publish it! Did you participate in summer research or a senior thesis? Publish it! Is there a part of your master's thesis you'd love to get out there but lack the funds? Publish it! *ideaFest Journal* is and always will be free-of-cost to publish and access online, and submissions are open to anyone in the Humboldt State Universe.

We are fortunate to be part of such a diverse academic community at Humboldt State University. While it may not be perfect, we do have the right, and at times responsibility, to use our inquisitive minds and written words to provoke thought and create change. I invite you all to dig deep, think critically, and get your voice heard. The authors in this issue are exemplary of this action. The next issue of *ideaFest Journal* will be published in 2020, what is sure to be a pivotal year in American history. The best way to fight anti-intellectualism is to come back a thousand voices strong. Whatever drives your search for knowledge, share it—show what our minds are really made of.

CAROLYN DELEVICH

Managing Editor, *ideaFest Journal*

THANKS TO OUR CONTRIBUTORS

Producing a journal takes a team and I can't thank enough the group of dedicated individuals that helped put this transformative journal together.

Thank you to **KACIE FLYNN** and HSU Sponsored Programs Foundation for your incredible support of student research on campus. You are the engine that makes *ideaFest* the crown jewel in empowering students and celebrating their achievements with the campus and the world.

Thank you to Managing Editor **CAROLYN DELEVICH**. Yet again you have put together a student campus research journal with content to rival any university publication, and with a design to rival any journal period. I will miss your quiet leadership, fierce intelligence, high ethics and accountability, and dedication to raise works to their highest potential (mine included). Thank you.

Thank you to Dean **CYRIL OBERLANDER** for your vision and commitment to celebrate original student research on campus. People see all the big things you do to support student success, but only because of the million little things you do is an endeavor like this possible.

Thank you to Professor **JANELLE ADSIT** for overseeing the first volume and laying the foundation upon which this volume stands. You provided all the guidance and support that made this edition possible.

Thank you to Professor **DAVID STACY** and his Spring 2019 **ENGL 318 RHETORIC FOR WRITERS STUDENTS** for providing peer review analyses on all submitted manuscripts. The students provided insightful reviews in the name of campus research, and themselves received a unique hands-on experience in this core element of academic research writing.

Thank you to the cohort of **HSU FACULTY**, **LECTURERS**, and **LIBRARIANS** that ensured that every article received a minimum of two peer-reviews. I have never worked with such an amazing community that is so dedicated to student success. You have provided these students with a challenge in scholarly communications and the rare opportunity to meet that challenge and publish their work in a peer-reviewed journal accessible around the world.

And a special thank you to all those who put yourself on the line by submitting an article. Whether published or not, you are all risk takers who strive to create change by sharing your voice with the world. Those who were not chosen for this issue, I expect to see your submissions for the next publication. Achievement does not come easy, but once accomplished, it is something to cherish.

KYLE MORGAN
Scholarly Communications Librarian
Humboldt State University Press

Preparing a Database of Extremely High Velocity Outflows in Quasars

Griffin Kowash[1*], Carla Quintero[1], Sean Haas[1], Paola Rodríguez Hidalgo[2]

ABSTRACT—Quasars are among the most distant and luminous objects in the universe. They form when matter spirals into the supermassive black hole at the center of a galaxy, creating a highly energetic disk of gas called an accretion disk. Our research focuses on outflows, which occur when a body of gas is ejected from the accretion disk at high speed. By examining how light from the quasar is absorbed by matter in an outflow, we can deduce the outflow's velocity relative to the quasar. A special subset of these events, termed extremely high velocity outflows (EHVOs), demonstrate outflows with speeds greater than 30,000 km/s, or 10% the speed of light. The goal of our research is to identify new instances of EHVOs, as there are currently only three confirmed cases in the published literature. Our examination of quasar spectra has yielded 40 new EHVO cases, increasing the number of confirmed cases by a factor of ~13. We are in the process of finishing an EHVO database, which includes quasar spectra and outflow parameters, and we will release it in late 2019 for the community to study the properties of these outflows, quasars, and their host galaxies in greater depth.

KEYWORDS—*astronomy, physics, astrophysics, quasars, active galactic nuclei*

INTRODUCTION—When quasars were first observed in the mid-20th century, their nature was a mystery. Superficially, they resemble stars, as both types of objects appear to us as point sources of light of similar brightness. As a result, they were given the name "quasi-stellar object," later abbreviated to "quasar." However, closer examination of the spectra reveals major discrepancies. First, the background radiation signature of a quasar, known as the continuum, appears to be distinct from that of a star; stellar spectra exhibit absorption lines superimposed on a blackbody spectrum. Second, quasar spectra contain broad emission lines, which are not present in stars. Finally, they display much higher cosmological redshifts than any intragalactic objects, placing most at a distance on the order of billions to tens of billions of light years from Earth. The current record holder for the most distant quasar, designated ttvft J1342 + 0928, is nearly 30 billion light years away.[1]

At first, researchers struggled to reconcile these observations. Given their great distance, these objects would need to have luminosities greater than that of most galaxies, but compacted into a far smaller space. Even nuclear fusion, the driving mechanism in the core of stars, was unable to account for this high energy density. Finally, in 1964 Zel'dovich proposed that clouds of gas falling under the force of gravity toward a supermassive black hole could liberate enough energy to explain the quasar phenomenon.[2] Although we now understand much more about their structure and behavior, quasars and their associated phenomena are still a very active area of research.

Our research focuses on a common phenomenon called an outflow, which occurs when a body of gas is ejected from a quasar's accretion disk at high speed. If an outflow intercepts the line of sight between us and the quasar, it absorbs some light from the continuum, leaving a distinctive signature in the spectrum. Not all spectra show evidence of outflows, but since our observation of

*Corresponding Author: gmk94@humboldt.edu
[1]Undergraduate Student
Department of Physics and Astronomy
Humboldt State University, Arcata, CA 95521
[2]Assistant Professor
Physical Sciences Division
University of Washington – Bothell, Bothell, WA 98011

Kowash, G., Quintero, C., Haas, S., Rodríguez Hidalgo, P. 2019. Preparing a database of extremely high velocity outflows in quasars. *ideaFest Journal: Interdisciplinary Journal of Creative Works & Research from Humboldt State University* 3: 6–9.

outflows depends on their intersection with our line of sight, it is reasonable to suspect that most if not all quasars exhibit outflows, while we perceive only those that have appropriate spatial positioning. Their potential ubiquity supports the idea that outflows play an important role in galactic evolution and star formation, in particular by expelling metal-rich gas with high kinetic energy into the interstellar medium.[3]

Our study focuses on a subset of outflows known as extremely high velocity outflows (EHVOs), which possess velocities greater than 0.1c (where c is the speed of light, approximately 3.00×10^8 m/s) with respect to the quasar rest frame. These outflows have received much less study than their lower-velocity counterparts, and there are currently only three confirmed cases of EHVO in the published literature.[4,5,6] Using quasar spectra from the Sloan Digital Sky Survey Data Release 9 (SDSS DR9), we identified 40 instances of EHVO, expanding the available data set by a factor of ~13.

METHODS—Our identification of EHVOs from quasar spectra revolves around the use of the carbon-IV ion (CIV; carbon three times ionized) as a marker. The signature of CIV appears in the spectrum in two ways: as an emission peak from ions in the broad emission line region, and as an absorption trough from ions in the outflow. If the quasar and the outflow were stationary with respect to each other, then the emission and absorption lines would appear at the same wavelength. In reality, outflow absorption tends to appear at a shorter wavelength than emission, as illustrated in FIG 1.

This wavelength offset indicates that the outflow is moving away from the quasar in our direction. Carbon atoms absorb photons at about 1,550 Å, but in the rest frame of the outflow, light emitted from the quasar is redshifted to a longer wavelength. As a result, photons absorbed by outflowing carbon must have a wavelength shorter than 1,550 Å in the quasar's rest frame. By comparing the observed wavelengths of the emission and absorption lines, we can calculate the blueshift of the outflow, and by extension its velocity with respect to the quasar. If this velocity is found to be greater than 30,000 km/s, it is evidence of an EHVO.

One complicating factor is that intervening material between Earth and the quasar may produce absorption features in the same wavelength range as our target CIV absorption. To distinguish between these effects, absorption features are categorized into two groups based on width: broad absorption lines (BALs) and narrow absorption lines (NALs). A number of factors can contribute to the broadening of absorption line profiles, for example Doppler broadening.[7] Individual atoms in an EHVO can have a wide range of velocities, and by extension different blueshifts with respect to the quasar. Atoms with speeds in the faster end of this range will produce absorption lines at a shorter wavelength, and vice versa. As a result, absorption occurs over a range of wavelengths. In contrast, a cloud of intervening matter has a more localized distribution of velocities, so it will produce much narrower absorption lines.

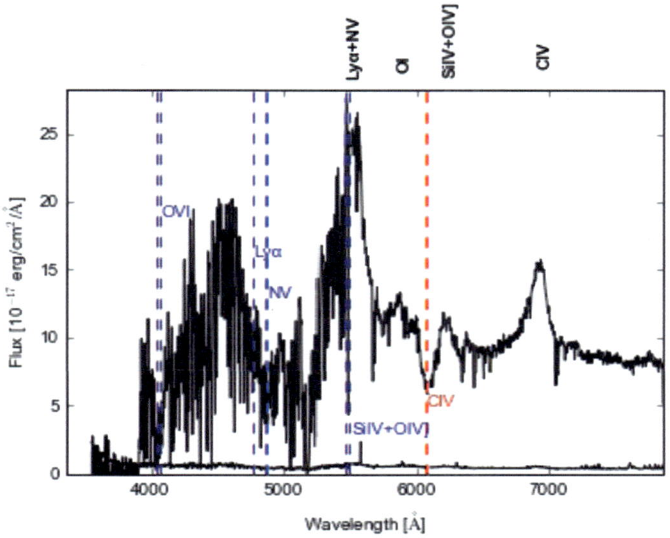

FIGURE 1. Example of a quasar spectrum with distinct CIV emission and absorption features caused by an EHVO. The absorption feature, marked by the red dashed line, appears at an observed wavelength nearly 1,000 Å shorter than the corresponding emission line at 7,000 Å, indicating that the absorbing matter is outflowing at a high velocity. At the top of the spectrum we indicate the location of the quasar emission lines. Other potential ions outflowing at the same speeds as the EHVO CIV are marked as dashed blue vertical lines.

It is worth noting that, although distinguishing between BALs and NALs can help identify the source of an absorption line, it is not an entirely decisive metric; there is evidence that some NALs are physically associated with the quasar, and may even represent outflows themselves.[8,9] Some models suggests that these outflowing NALs may simply result from viewing a standard outflow at a shallower angle, but more research is needed to clarify their true nature.[10]

For our analysis, we used a set of 87,822 quasar spectra obtained from the SDSS DR9.[11] Thresholds were placed both on the signal-to-noise ratio and the cosmological redshift; the signal-to-noise threshold helped ensure that signs of EHVO would not be produced by random signal noise, while the redshift threshold ensured that our region of interest fell within the SDSS spectrograph's range of wavelength. Further details on these cutoffs can be found in our previous publications.[12,13] After applying these two filters, our data set was reduced to 6,760 spectra.

Next, the spectra were normalized before analyzing for emission and absorption features. These features appear in the spectra as regions where the flux rises above or falls below the continuum of the quasar, respectively. However, as shown in FIG 2A, the continuum has a non-zero slope in the UV-optical region, making it difficult to recognize and quantify emission and absorption. Adjusting the data to give the continuum a constant value of one creates a common baseline against which flux values can be compared, as illustrated by the horizontal line in FIG 2B.

This region of the continuum is well approximated by a power law,[14] and by selecting anchor points in the spectrum where there is known to be little contaminating emission or absorption, a curve fit can be calculated for use in normalization in a systematic way. Finally, we removed spectra that were either missing data in the wavelength region of interest or had too complex of a continuum, so we ended up with a parent sample of 6,740 normalized quasar spectra.

Once the spectra were normalized, potential EHVO absorption features were flagged (and if the identification as CIV was correct, quantified) using a modified form of the Balnicity Index (BI), given by the following integral:

$$BI = \int_{30000}^{60000} \left[1 - \frac{f(v)}{0.9}\right] C\, dv$$

where *f(v)* is the normalized flux as a function of velocity, and C is a parameter whose value is either zero or one.[14] If the quantity in brackets remains positive over an interval longer than a specified threshold, the value of C becomes one; otherwise it is given a value of zero. A larger threshold interval means that only broader absorption features are summed under the integral. We required this threshold to be 2,000 km/s to remove potential blends of narrow lines more easily (as discussed below). Thus, by selecting an appropriate threshold and finding all spectra with BI > 0, quasars displaying BALs can be filtered from the data set. In the original definition of the BI introduced in Weymann et al. (1991),[15] the integral bounds extend from 3,000 km/s to 25,000 km/s; however, this interval excludes EHVOs, which have velocities greater than 30,000 km/s, so it was necessary to adjust the bounds.

Only a subset of these BALs will be due to CIV absorption. Broad absorption lines occurring at wavelengths shorter than 1,400 Å may also be caused by SiIV (triply ionized silicon), CII (singly ionized carbon), or OI (unionized oxygen). However, if this is the case, CIV absorption should also be visible, outflowing at similar speeds (or in other words, at the corresponding redshift), since

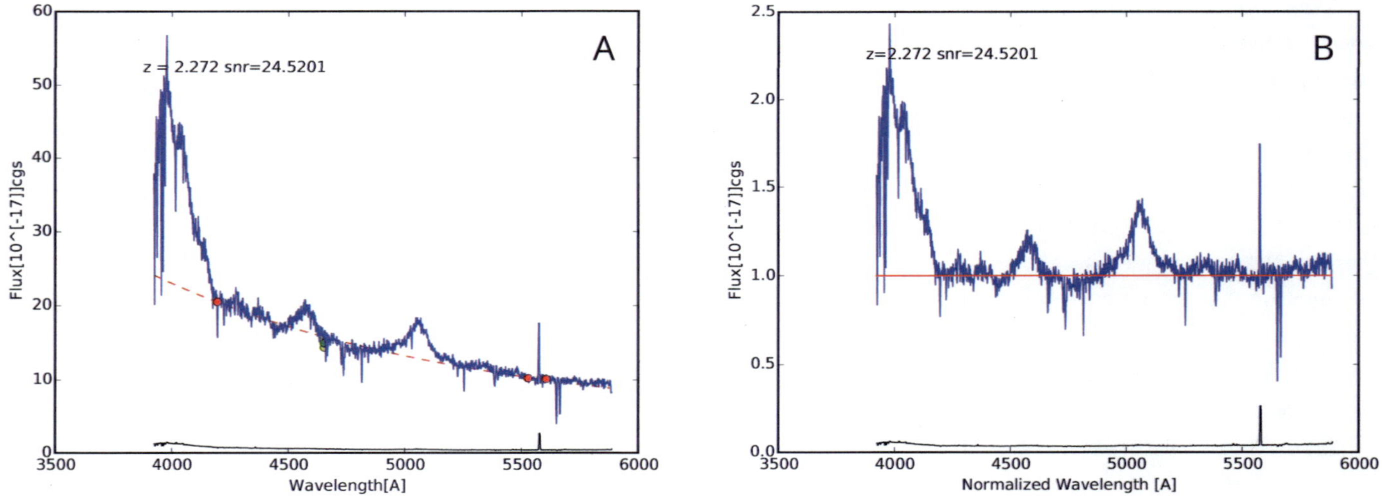

FIGURE 2. (A) Quasar spectrum prior to normalization. The dashed diagonal line shows the continuum of the quasar, and red circles denote anchor points. (B) Spectrum following normalization. Regions above and below the horizontal line indicate emission and absorption, respectively.

CIV is the predominant absorbing ion observed in quasar spectra. Also, in SDSS spectra BAL-like features can be due to the blending of multiple narrow absorption lines, which could correspond to intervening matter as discussed above. These features can be distinguished visually from true BALs by the slope of the trough boundary. If an absorption feature is the result of many superimposed narrow absorption lines, the boundary will tend to have a steep slope, while a true BAL tapers off more gradually. In addition, the presence of spikes at the bottom of the trough suggests it is probably composed of narrow absorption lines. Therefore, visual inspection is used to confirm cases of EHVO out of the potential flagged cases. Additional confirmation that a BAL is due to CIV absorption comes from the presence of NV (nitrogen four times ionized) and OVI (oxygen five times ionized) absorption at the same redshift.[17]

RESULTS—Following the procedure outlined above yielded 40 confirmed cases of EHVOs, such as the one displayed in FIG 1, from the parent sample of 6,740 spectra. Our results, including spectral plots and absorption measurements (such as BI, velocity of the outflow, and depth of the absorption trough) for all 40 EHVO cases, will be released in 2019 through a publicly accessible online database, allowing other groups to begin a more in-depth analysis of their properties.

CONCLUSIONS—Extremely high velocity outflows are a highly understudied phenomenon associated with quasars, and they may play an important role in star formation and galactic evolution. Our search of the SDSS DR9 data set yielded 40 confirmed EHVO cases, expanding the current set of 3 published cases by a factor of ~13. Our upcoming work includes publishing our results in an online database in 2019, along with an accompanying journal article that includes a more comprehensive analysis of the properties of our sample, allowing the astronomy community to begin studying the properties of these phenomena in unprecedented depth.

REFERENCES

[1] Bañados, E. et al. 2018. An 800-million-solar-mass black hole in a significantly neutral Universe at a redshift of 7.5. *Nature* 553: 473–478.

[2] Zel'dovich, Y. B. 1964. The fate of a star and the evolution of gravitational energy upon accretion. *Soviet Physics–Doklady* 9: 195–197.

[3] Di Matteo, T., Springel, V., Hernquist, L. 2005. Energy input from quasars regulates the growth and activity of black holes and their host galaxies. *Nature* 433: 604–607.

[4] Hamann, F., Barlow, T. A., Beaver, E. A., Burbidge, E. M., Cohen, R. D. 1995. NE VIII Lambda 774 and Time Variable Associated Absorption in the QSO UM 675. *The Astrophysical Journal* 443: 606–616.

[5] Jannuzi, B. T. et al. 1996. The Hubble Space Telescope quasar absorption line key project: The unusual absorption line system in the spectrum of PG 2302+029—Ejected or intervening? *The Astrophysical Journal* 470: L11–L14.

[6] Rogerson, J.A., Hall, P. B., Rodríguez Hidalgo, P., Pirkola, P., Brandt, W. N., Ak, N. F. 2016. Multi-epoch observations of extremely high-velocity emergent broad absorption. *Monthly Notices of the Royal Astronomical Society* 457: 405–420.

[7] Robinson, K. 2007. *Spectroscopy: The Key to the Stars*. Springer, New York, NY.

[8] Misawa, T., Charlton, J. C., Eracleous, M., Ganguly, R., Tytler, D., Kirkman, D., Suzuki, N., Lubin, D. 2007. A census of intrinsic narrow absorption lines in the spectra of quasars at Z = 2–4. *The Astrophysical Journal Supplement Series* 171: 1–28.

[9] Nestor, D., Hamann, F., Rodríguez Hidalgo, P. 2008. The quasar-frame velocity distribution of narrow CIV absorbers. *Monthly Notices of the Royal Astronomical Society* 386: 2055–2064.

[10] Elvis, M. 2000. A Structure for Quasars. *The Astrophysical Journal* 545: 63–76.

[11] Pâris, I. et al. 2012. The Sloan Digital Sky Survey quasar catalog: Ninth data release." *Astronomy & Astrophysics* 548: A66.

[12] Quintero, C., Haas, S., Rodríguez Hidalgo, P. 2018. How many quasars have extremely high velocity outflows? *ideaFest: Interdisciplinary Journal of Creative Works and Research from Humboldt State University* 2: 23–26.

[13] Khatri, A. M. 2015. Final Report—Extremely High Velocity Outflows. University of Toronto.

[14] Natali, F., Giallongo, E., Cristiani, S., La Franca, F. 1998. The optical-ultraviolet continuum of a sample of QSOs. *The Astronomical Journal* 115: 397–404.

[15] Weymann, R., Morris, S. L., Foltz, C. B., Hewett, P. C. 1991. Comparisons of the emission-line and continuum properties of broad absorption line and normal quasi-stellar objects." *The Astrophysical Journal* 373: 23–53.

[16] Rodríguez Hidalgo, P., Hamann, F., Hall, P. 2011. The extremely high velocity outflow in quasar PG0935+417. *Monthly Notices of the Royal Astronomical Society* 411: 247–259.

[17] Rodríguez Hidalgo, P. 2009. High Velocity Outflows in Quasars. PhD Dissertation, University of Florida.

Searching for Trends in Atmospheric Compositions of Extrasolar Planets

Kassandra Weber[1*], Paola Rodríguez Hidalgo[2], Adam Turk[1], Troy Maloney[1], Stephen Kane[3]

ABSTRACT—Since the first exoplanet was discovered decades ago, there has been a rapid evolution of the study of planets found beyond our solar system. A considerable amount of data has been collected on the nearly 3,838 confirmed exoplanets found to date. Recent findings regarding transmission spectroscopy, a method that measures a planet's upper atmosphere to determine its composition, have been published on a limited number of exoplanets. The aim of our work was to gather existing data on atmospheric planetary composition and search for potential trends in relation to exoplanets' orbital and physical properties. Due to their short periods and thicker atmospheres, hot Jupiter-type planets wereour first target population. Out of 78 cases with periods shorter than three Earth days and radii larger than 1 R_J (Jupiter radius), we found previously-published data on the atmospheres of 15 hot Jupiters. Only eight cases had an overlapping wavelength range that allowed comparisons: 4,800–9,000 Å. Within that range only one exoplanet shows absorption in their atmospheric data. We report our findings on this set, which will be publicly available on the Habitable Zone Gallery, and our future work plans for expanding the dataset.

KEYWORDS—*exoplanets, atmospheric composition, transmission spectroscopy*

INTRODUCTION—In the past two decades, the search for life on other planets has increased exponentially. Most of the recent developments in this new area of science have revolved around the observations of exoplanets, which are planets that are in orbit around a star outside of our solar system. To date, 3,838 exoplanets have been confirmed; a current list of detected extrasolar planets can be found on NASA's Exoplanet Archive.[1]

Exoplanets are detected using several techniques. One of these techniques is the transit method, which is done by observing the planets as they pass in front of their host star as seen from Earth.[2] This method has been one of the most successful methods to date—2,998 exoplanets have been discovered using the transit method.[3]

The easiest planets to detect are those called hot Jupiters. While we don't have any example of hot Jupiters in our own solar system, many have been detected orbiting other stars. They are easier to detect due to their large mass and close proximity to the parent stars; because of their proximity, they have shorter orbital periods and more frequent transits.

The transit method also provides us with other stellar properties of each exoplanet when used in combination with transmission spectroscopy. Each transit produces two dips in light—once as the planet passes in front of its star as seen from Earth, and again as it passes behind the star (**FIG 1**). While in front of the star, in primary transit, we get a glimpse of the starlight combined with the light shining through the planet's atmosphere. As the planet passes behind its star, in its secondary eclipse, or occultation, only the pure starlight is seen. The difference between the spectral data observed from the secondary eclipse and primary transit reveals the spectral features and emissions of the exoplanet itself. Researchers analyze this using spectroscopy, the study of visible light dispersed according to its wavelength. If the star size is known, this method also informs us of the radius of the exoplanet using the planet to star area ratio (**FIG 1**).

*Corresponding Author: kaw726@humboldt.edu
[1]Undergraduate Student
Department of Physics and Astronomy
Humboldt State University, Arcata, CA 95521
[2]Assistant Professor
Physical Sciences Division
University of Washington – Bothell, Bothell, WA 98011
[3]Associate Professor
Department of Earth and Planetary Sciences
University of California – Riverside, Riverside, CA 92521

Weber, K., Rodríguez Hidalgo, P., Turk, A., Maloney, T., Kane, S. 2019. Searching for trends in atmospheric compositions of extrasolar planets. *ideaFest Journal: Interdisciplinary Journal of Creative Works & Research from Humboldt State University* 3: 10–14.

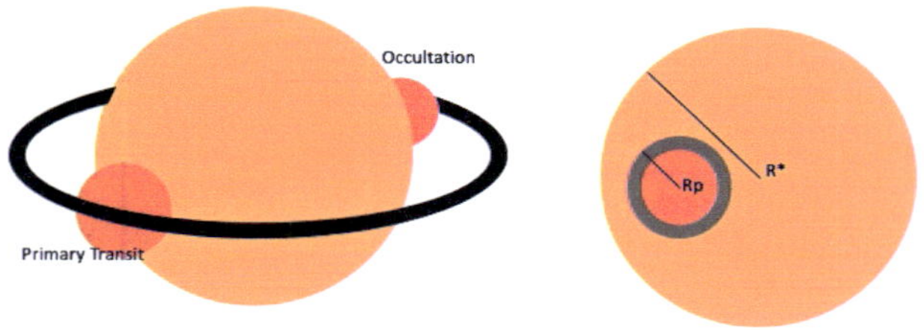

FIGURE 1. A depiction of a planet in transit around its parent star in an arbitrary eccentric orbit, or an orbit that is a deviation of a circle (left). While it passes between the star and the observer, it is in primary transit. When the planet transits behind the star relative to the observer, it is then in occultation, or secondary eclipse. As the planet lies in front of the star relative to us, part of the stellar light passes through the exoplanet's atmosphere (as seen shaded in grey on the right). This light is absorbed by the ions and molecules present, giving us information about the atmospheric composition. The size of the atmosphere and the exoplanet have been exaggerated for clarity.

In this paper we present a study of potential trends between atmospheric composition and physical properties of exoplanets. To do so, we used Habitable Zone Gallery data and published spectroscopic data. The Habitable Zone Gallery, a website by Stephen Kane and Dawn Gelino,[4] was created as a database for exoplanetary systems for which the orbital parameters have been measured. The gallery includes stellar and planetary parameters extracted from the Exoplanet Data Explorer and calculations of the amount of time a planet spends in its habitable zone.[5,6] The Habitable Zone Gallery does not currently provide spectroscopic information of the exoplanets' atmosphere. Our work will help supplement this database so that this information will be available to the astronomy community.

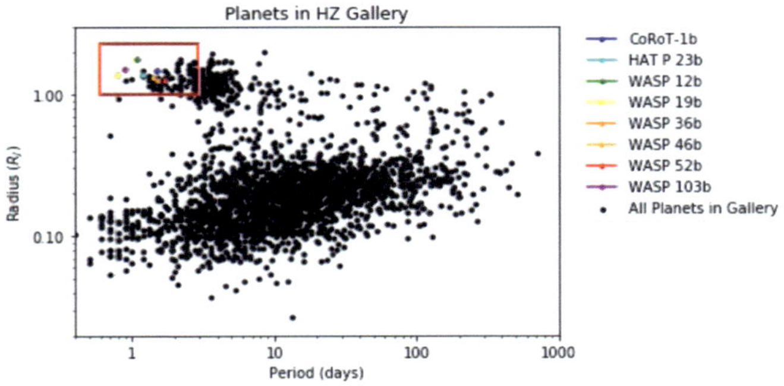

FIGURE 2. This figure shows the orbital periods (in Earth days) versus radius (Jupiter radius, R_J) of all known exoplanets available in the Habitable Zone Gallery. The red box encompasses the planets within the selected parameter space: 1–2 R_J and periods between 1–3 Earth days, and the eight planets we analyzed (see Results section).

METHODS—We began our study by downloading the most current list of exoplanets as of February 2017 from the Habitable Zone Gallery, which included 3,833 exoplanets.

In order to narrow down the amount of planets for our search, we decided to focus on hot Jupiter-type exoplanets, which are physically some of the easiest planets to detect due to their massive size and thick atmosphere, as well as their short orbital periods. In the early stages of our work, we selected exoplanets with radii > 1 R_J (Jupiter radius) and orbital periods of less than two Earth days.[6] As the study progressed, we expanded upon previous work by modifying our cutoff for the orbital period to three Earth days (FIG 3). We wrote our own Python script to plot and select our targets. We found 78 exoplanets within this region. We have included data on these exoplanets in TABLE S1.

RESULTS—Once we identified the list of exoplanets within the parameter space of interest, we searched the literature for published spectral data on these exoplanets. Out of the 78 exoplanets in our parent sample, only 15 had available data on their atmospheric composition. In addition, the planets in this subset had been observed in various wavelength ranges. To be able to establish comparisons we needed to select an overlapping wavelength region. We found that eight of them had a large overlap within their spectral wavelength range: 480–900 nm.[1,7,8,9,10,11,12,13,14,15]

FIG 3 shows the reproduced spectra of these eight exoplanets. Exoplanets' spectra are shown as R_p/R^* vs wavelength; R_p/R^* is the ratio between the observed radius of the planet and the observed radius of its star. This ratio is proportional to the stellar flux (EQ 1), or the energy radiated by the star as measured here on Earth, which is different at different wavelengths.

This is a useful tool in analyzing correlations since it takes into account the difference in starlight per each system.

$$\sqrt{\frac{\Delta F}{F_s}} = \frac{R_p}{R_s} \qquad \text{(Eq 1)}$$

The spectra of each exoplanet seems to be relatively

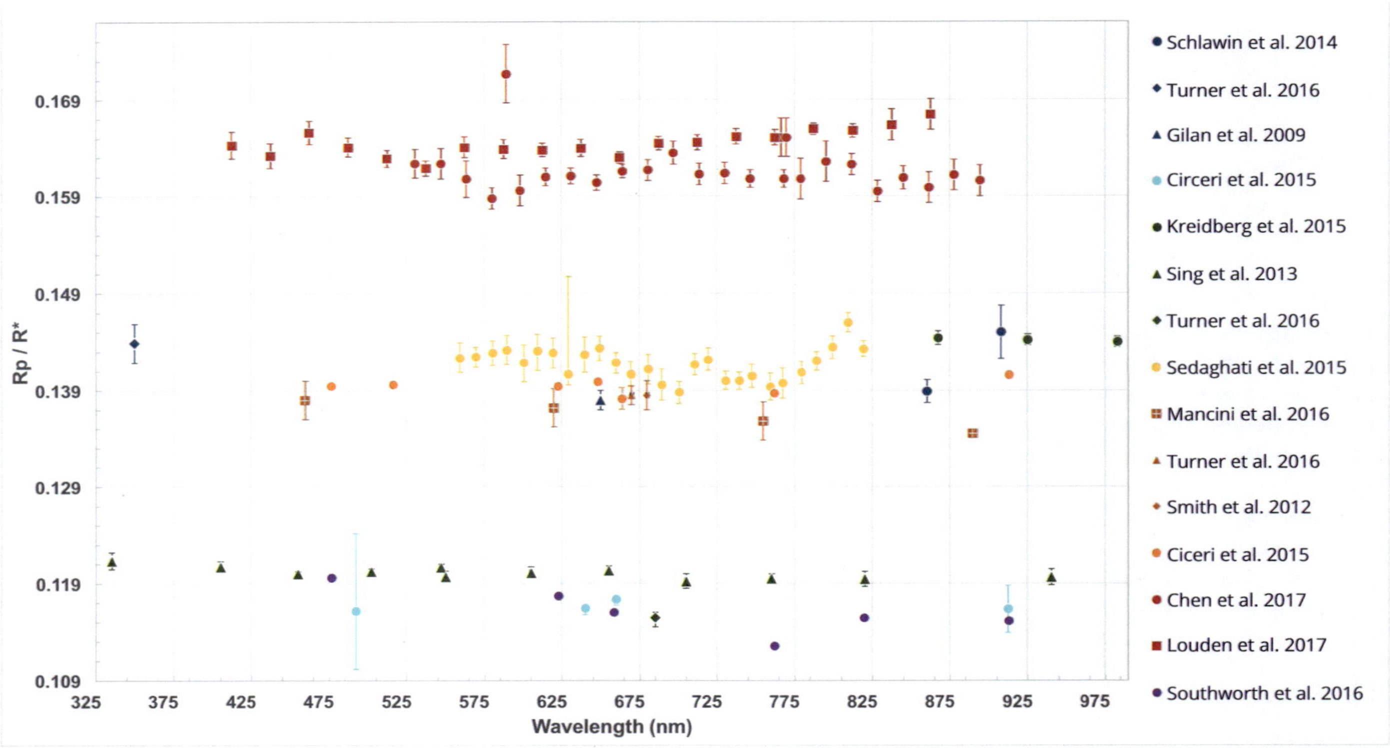

FIGURE 3. Reproduced spectra of the eight exoplanets that had published data on the selected common wavelength range (480–900 nm): CoRoT-1b,[11,15] HAT-P 23b,[8] WASP 103b,[14] WASP 36b,[1,15] WASP 46b,[8] WASP 12b,[13,15] WASP 19b,[1,9,12] and WASP 52b.[7,10] Here we can see a peak in the spectral data of WASP-52b at 589 nm. Once more literature is published and confirmations of peak absorptions are made within each planet's spectral data, we might be able to clearly isolate other peaks amongst the dataset and analyze them accordingly.

flat across all wavelengths, with the exception of the spectral data of WASP-52b, where there is an indication of absorption at approximately 589 nm (**FIG 3**). Consequently, with only one exoplanet (WASP-52b) that shows absorption in their spectral data, we cannot explore any trends between physical characteristics, orbital properties, and atmospheric composition. Although this is the case, there are multiple exoplanets with radii and orbital periods similar to WASP-52b but without published literature (**FIG 4**).

WASP-52b physical characteristics are fairly notable compared to the other planets within our set (**FIG 5**). It has the largest period of 1.7 days and the highest R_p/R^* in our dataset. It has the smallest radius of 1.27 R_J, the smallest mass, and the lowest temperature of all the studied cases. Planets with features similar to those of WASP-52b might produce similar spectral data, but this will require further investigation once more literature becomes available.

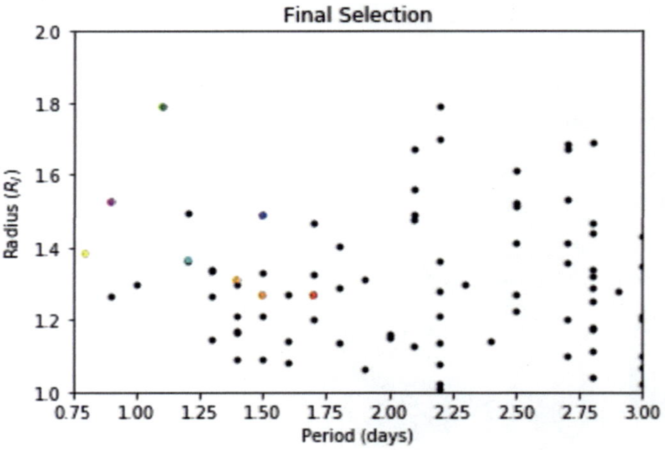

FIGURE 4. This plot includes the 78 planets within our range of parameters. The eight planets with overlapping spectral data are highlighted.

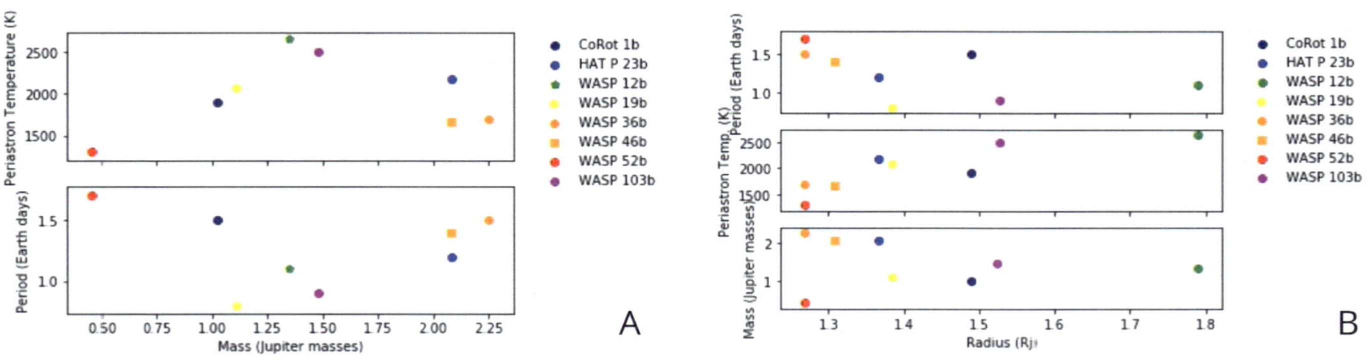

FIGURE 5. (A) The eight exoplanet cases with their Jupiter radii versus Jupiter mass, Periastron surface temperatures, and periods. WASP-52b notably lies on the far left with the smallest mass and surface temperature. (B) The eight exoplanet cases with their Jupiter masses versus periods in Earth days and Periastron surface temperatures. WASP-52b also has the longest period and smallest radius.

CONCLUSION—Though the study of exoplanets is fairly new, in the past two decades it has produced an incredible amount of data, allowing researchers to expand upon our knowledge of these distant worlds. We have initiated a project to search for potential correlations between the physical characteristics of exoplanets and their atmospheric composition. The hot-Jupiter-type planets we targeted had orbital periods of less than three Earth days and large radii ($1 < R_J < 2$). From these 78 planets, our search of available published atmospheric data narrowed down 15 planets, eight of which show spectra within the same wavelength region (480–900 nm) to allow comparisons among the spectra. Out of those, only one spectrum shows absorption, so no trends could be explored.

Our parent sample was largely reduced (from 78 to 8) due to the lack of available spectra and overlapping spectral regions, but we expect that the current growth of atmospheric exoplanet research will result in more cases to add to our study in the near future. The reproduced spectral data will be available soon in the Habitable Zone Gallery website to provide access for the community to carry out these types of studies.

REFERENCES

[1] Mancini, L., Kemmer, J., Southworth, J., Bott, K., Mollière, P., Ciceri, S., Chen, G., Henning, T. 2016. An optical transmission spectrum of the giant planet WASP-36b." *Monthly Notices of the Royal Astronomical Society* 459: 1393–1402.

[2] Seager, S. 2010. *Exoplanet Atmospheres: Physical Processes*. Princeton University Press, Princeton, NJ.

[3] NASA Exoplanet Archive. 2017. Accessed December 10, 2017. http://exoplanetarchive.ipac.caltech.edu/.

[4] Kane, S. R., Gelino, D. M. 2012. The Habitable Zone Gallery. *Publications of the Astronomical Society of the Pacific* 124: 323–328.

[5] Fakhouri, O. "Exoplanet Data Explorer. "Exoplanet Orbit Database. Accessed December 10, 2017. http://exoplanets.org/.

[6] Maloney, T. 2017. A catalog of exoplanet transmission spectroscopy data for use in atmospheric studies. *Proceedings of the National Conference on Undergraduate Research 2017*, 7 Apr. 2017.

[7] Chen, G., Pallé, E., Nortmann, L., Murgas, F., Parviainen, H., Nowak, G. 2017. The GTC exoplanet transit spectroscopy survey. IV. Detection of sodium in WASP-52b's cloudy atmosphere. *Astronomy & Astrophysics* 600: L11.

[8] Ciceri, S., et al. 2015. Physical properties of the HAT-P-23 and WASP-48 planetary systems from multi-colour photometry. *Astronomy & Astrophysics* 577: A54.

[9] Huitson, C. M., et al. 2013. An *HST* optical-to-near-IR transmission spectrum of the hot Jupiter WASP-19b: Detection of atmospheric water and likely absence of TiO." *Monthly Notices of the Royal Astronomical Society* 434: 3252–3274.

[10] Louden, T., Wheatley, P. J., Irwin, P. G. J., Kirk, J., Skillen, I. 2017. A precise optical transmission spectrum of the inflated exoplanet WASP-52b. *Monthly Notices of the Royal Astronomical Society* 470: 742–754.

[11] Schlawin, E., Zhao, M., Teske, J. K., Herter, T. 2014. A 0.8–2.4 μm transmission spectrum of the hot jupiter CoRoT-1b. *The Astrophysical Journal* 783: 5.

[12] Sedaghati, E., Boffin, H. M. J., MacDonald, R. J., Gandhi, S., Madhusudhan, N., Gibson, N. P., Oshagh, M., Claret, A., Rauer, H. 2017. Detection of titanium oxide in the atmosphere of a hot Jupiter. *Nature* 549: 238–241.

[13] Sing, D. K., et al. *HST* hot-jupiter transmissionspectral survey: Evidence for aerosols and lack of TiO in the atmosphere of WASP-12b. *Monthly Notices of the Royal Astronomical Society* 436: 2956–2973.

[14] Southworth, J., et al. 2014. High-precision photometry by telescope defocusing. VII. The ultrashort period planet WASP-103. *Monthly Notices of the Royal Astronomical Society* 447: 711–721.

[15] Turner, J. D., et al. 2017. Investigating the physical properties of transiting hot Jupiters with the 1.5-m Kuiper Telescope. *Monthly Notices of the Royal Astronomical Society* 472: 3871–3886.

SUPPLEMENTARY DATA

TABLE S1. List of exoplanets with radii $1 < R_J < 2$ and periods between 1 and 3 Earth days found in the Habitable Zone Gallery. An asterisk "*" indicates the exoplanets that have published literature on their atmospheric data [c18] .

Name	Period (Earth day)	Radius (R_J)	Avg. Temp (K)
CoRoT-12b	2.8	1.44	1445.35
CoRoT-14b	1.5	1.09	1951
CoRoT-18b	1.9	1.31	1526.75
CoRoT-1b	1.5	1.49	1899.4
CoRoT-2b	1.7	1.466	1537
HAT-P-13b	2.9	1.281	1647.65
HAT-P-16b	2.8	1.289	1625.15
HAT-P-23b	1.2	1.368	2058.95
HAT-P-30b	2.8	1.34	1637.35
HAT-P-32b	2.2	1.789	1785
HAT-P-36b	1.3	1.264	1820.95
HAT-P-37b	2.8	1.178	1276.75
HAT-P-41b	2.7	1.685	1937
HAT-P-49b	2.7	1.413	2127.8
HAT-P-56b	2.8	1.466	1839
HAT-P-57b	2.5	1.412	2197.7
HAT-P-5b	2.8	1.254	1537.1
HAT-P-7b	2.2	1.363	2225.2
HATS-14b	2.8	1.039	1274.6
HATS-2b	1.4	1.168	1575.3
HATS-4b	2.2	1.02	1385.7
HATS-9b	1.9	1.065	1816.4
HD 189733b	2.2	1.138	1200.1
KELT-3b	2.7	1.358	1821.3
KELT-7b	2.7	1.533	2049.9
Kepler-17b	1.5	1.33	1744.9
Kepler-412b	1.7	1.325	1828.15
Kepler-423b	2.7	1.2	1412.5
Kepler-670b	2.8	1.176	1388.6
Kepler-686b	1.6	1.084	1614.9
Kepler-718b	2.1	1.477	1862.2
Kepler-785b	2.0	1.162	1146
Kepler-840b	2.5	1.523	1538.5
Kepler-854b	2.1	1.492	1812.9
KOI-13b	1.8	1.406	2607
OGLE2-TR-L9b	2.5	1.61	2033.8
OGLE-TR-113b	1.4	1.093	1342.8
OGLE-TR-132b	1.7	1.2	1974.5
OGLE-TR-56b	1.2	1.363	2206
Qatar-1b	1.4	1.164	1389
Qatar-2b	1.3	1.144	1289.4
TrES-2b	2.5	1.224	1497.6
TrES-3b	1.3	1.336	1628.6
TrES-5b	1.5	1.209	1481.8
WASP-100b	2.8	1.69	2199.9
WASP-103b	0.9	1.528	2504.4
WASP-104b	1.8	1.137	1516.1
WASP-12b	1.1	1.79	2585.75
WASP-135b	1.4	1.3	1712.3
WASP-14b	2.2	1.281	1869.2
WASP-18b	0.9	1.67	2397.75
WASP-19b	0.8	1.386	2065.65
WASP-1b	2.5	1.516	1848.7
WASP-24b	2.3	1.3	1768.3
WASP-26b	2.8	1.32	1412.3
WASP-2b	2.2	1.077	1299.4
WASP-32b	2.7	1.1	1564.4
WASP-33b	1.2	1.497	2673.2
WASP-36b	1.5	1.269	1699.1
WASP-3b	1.8	1.29	1990.1
WASP-44b	2.4	1.14	1347.4
WASP-46b	1.4	1.31	1658
WASP-48b	2.1	1.67	2033.2
WASP-49b	2.8	1.115	1370.2
WASP-4b	1.3	1.341	1670.2
WASP-50b	2.0	1.153	1392.5
WASP-52b	1.7	1.27	1300.8
WASP-5b	1.6	1.14	1741.4
WASP-64b	1.6	1.271	1690.1
WASP-72b	2.2	1.01	2060.4
WASP-74b	2.1	1.56	1922.3
WASP-75b	2.5	1.27	1704.5
WASP-77Ab	1.4	1.21	1670.9
WASP-78b	2.2	1.7	2294.7
WASP-82b	2.7	1.67	2178.8
WASP-95b	2.2	1.21	1617.4
WASP-97b	2.1	1.13	1539.9
WTS-2b	1	1.3	1543.9

Graduate Student Perspectives on Scale and Hierarchy in Ecology

Carolyn Delevich[1], Adrian Macedo[1], Claire Nasr[2], Adrien Bouissou[1], Sabrina Horrack[1], Johnny Roche[2], Wesley Hull[2], Joseph Saler[1], Stacie Nunes[1], Vladimir Bonilla[1], Matthew Reilly[3]*

INTRODUCTION—Scale and hierarchy are unifying themes that span a broad range of ecological disciplines (Levin 1992). These themes received considerable attention and went through major conceptual developments in the 1980s and early 1990s, when advances in computing enabled theoretical development through mathematical modeling. Scale refers to the spatial or temporal domain at which a study takes place and includes two major components. The first component, grain, is the minimum resolution of an observation, while the second component, extent, is the total area or time span covered by all of the observations. Hierarchy refers to the various levels at which ecological entities are organized, ranging from genes and organisms to ecosystems and biomes, and may be observed across a range of spatial or temporal scales.

Scale and hierarchy offer a framework for organizing the complexity that characterizes ecological patterns, processes, and dynamics. Underlying the appeal of scale and hierarchy to ecologists is the potential to enhance a mechanistic understanding of how patterns manifested at one scale or level of organization can be explained by processes occurring at another scale or level of organization. Such an understanding can reduce ecological complexity by boiling down the multitude of processes and interactions into a few measurable and meaningful variables. The appeal and allure of scale and hierarchy are still as pervasive now as they were a few decades ago. However, questions remain on how these concepts have developed in both application and, perhaps more importantly, in their intellectual accessibility to early career ecologist in graduate school.

We participated in a semester-long graduate seminar that focused on the development and application of scale and hierarchy in ecology. Our research interests and experiences span a variety of ecological disciplines and range of spatial and temporal domains (FIG 1). They also reflect our status as early career scientists and the

*Corresponding Author: matthew.reilly@humboldt.edu
[1]Graduate Student
Department of Biological Sciences
Humboldt State University, Arcata, CA 95521
[2]Graduate Student
Department of Wildlife
Humboldt State University, Arcata, CA 95521
[3]Postdoctoral Fellow
Department of Biological Sciences
Humboldt State University, Arcata, CA 95521

Delevich, C., Macedo, A., Nasr, C., Bouissou, A., Horrack, S., Roche, J., Hull, W., Saler, J., Nunes, S., Bonilla, V., Reilly, M. 2019. *ideaFest Journal: Interdisciplinary Journal of Creative Works & Research from Humboldt State University* 3: 15–23.

limitations imposed by two-year studies and short funding cycles. Despite this, our interests and educational experiences collectively provide a much broader basis with which to assess recent advances and applications of scale and hierarchy to ecological research.

Collectively, we developed a list including some of the major themes and concepts in scale and hierarchy after reading and discussing seminal works on topics (Allen & Hoekstra 1992; Levin 1992; O'Neil & King 1996). We reduced this list to a manageable number of terms that are central to understanding concepts of scale and hierarchy in an ecological context (TABLE 1, TABLE S1). These terms generally apply to integrating or predicting across scales of hierarchical levels (e.g., cross-scale interactions, aggregation, emergent properties), but also include ecological properties relating to stability and equilibrium dynamics (e.g., resistance, resilience). We then performed literature reviews using these terms in conjunction with four ecological disciplines reflecting our collective research and intellectual experiences.

Our synthesis focuses on four distinct ecological disciplines that span marine and terrestrial systems, includes plants, animals, and fungi as organisms of interest, and even integrates an evolutionary perspective. Our major objective was to synthesize existing literature in our own respective ecological disciplines, then compare among disciplines. Given time constraints, we acknowledge that literature reviews may be far from exhaustive. However, we provide a common framework for synthesizing the collective knowledge on scale and hierarchy, as well as how it has developed over the last three decades.

SCALE AND HIERARCHY IN THE MYCORRHIZAL SYMBIOSIS—The mycorrhizal symbiosis is arguably one of the most important symbioses to life on land. Behind the scenes of the emergence of vascular plants onto land some 450 million years ago (Morris et al. 2018), mycorrhizal fungi were providing essential nutrients to plants that were slowly figuring out how to live on land. In present times, the plant-fungal mutualism remains cosmopolitan, forming in 80% of land plants worldwide (Wang & Qiu 2006). In this mutualism, fungi exchange water and soil nutrients for carbohydrates produced by the plant through photosynthesis (Smith & Read 1996). The mutualism presents itself in many forms, including arbuscular mycorrhizae, which penetrate host root cells, ectomycorrhizae, which operate through extracellular contact, and mycorrhizae that are specific to ericoid plants and to orchids (Smith & Read 1996). This section will focus particularly on ectomycorrhizae, which wrap around their host roots and share resources via cell-to-cell contact with the outer cells of the host roots. The mutualism is pervasive and is undeniably important to the ecology of land plants and ectomycorrhizal (ECM) fungi alike.

Given how prevalent the ECM symbiosis is phylogenetically and geographically, it is unsurprising that scientists have homed in on its ecological scale. While the symbiosis itself forms at the cellular level, emergent properties of the symbiosis can have consequences for populations, communities, ecosystems, and even global-scale processes. The symbiosis has incredible importance on the individual plant-fungus level—most plants and fungi that form the symbiosis do so obligately and cannot survive alone (Smith & Read 1996). On an ecosystem scale, ECM fungi are responsible for replenishing nitrogen into the soil environment and helping to reabsorb nutrients into plants; they are pivotal to nutrient cycling in forest ecosystems (Fogel 1980).

Many studies of the interactions between plants and their ECM fungi rely on laboratory-based manipulation. The feasibility of conducting laboratory-based studies becomes more and more difficult when we are looking

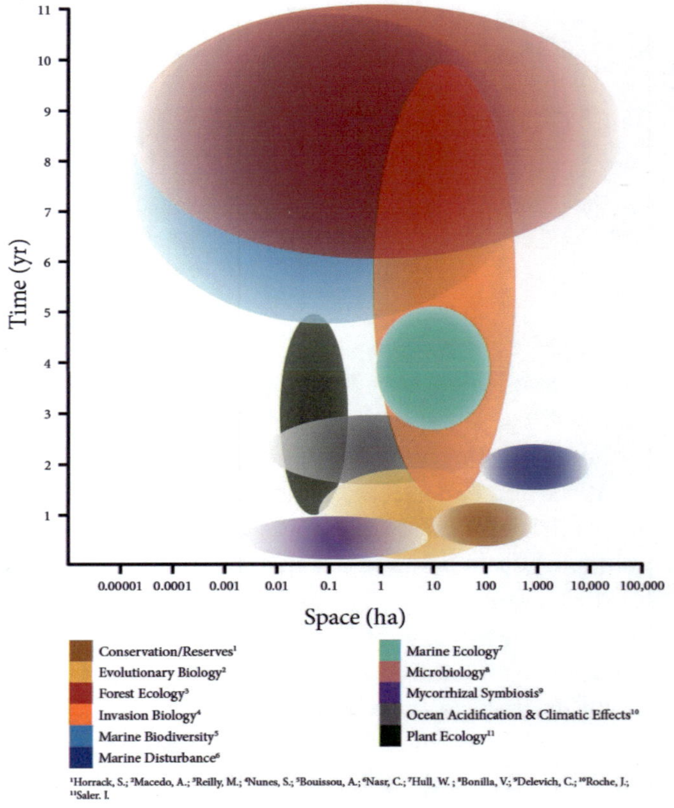

FIGURE 1. A space-time diagram displaying the various areas of focus of master's student thesis projects from the Department of Wildlife and the Department of Biological Sciences at Humboldt State University.

to answer large-scale questions. Novel techniques in the field-based study of the ECM symbiosis are quickly emerging, allowing us to begin to integrate across scales using empirical techniques. Perhaps one of the biggest breakthroughs in larger-scale field studies is the use of isotope analysis in studying common mycorrhizal networks (CMNs). Common mycorrhizal networks are underground systems whereby the fungal organ (mycelium) connects roots of different trees, allowing a pathway for sharing soil nutrients, water, and photosynthates. While the concept of CMNs is nothing new, developments in isotope analysis have allowed us to look deeper into these connections and see that they link not just related trees, but entire forest communities through source-sink relationships (Simard et al. 1997). Within these networks, fungi can preferentially allocate their nutrients to plants that provide more photosynthates, elucidating an element of competition in these systems (Fellbaum et al. 2014). By continuing to bolster techniques for field-based experimentation and observation, we will be able to fill in

TABLE 1. List of the major concepts and terms related to scale and hierarchy and the prevalence of each in different ecological disciplines. Cells are coded with qualitative ratings on the prevalence of each in the respective discipline. Darker shading indicates less term usage and letter codes are as follows: R = Rare, O = Occasional, C = Common, U = Ubiquitous.

Concepts and Terms	Terrestrial Disturbance	Evolutionary Biology	Conservation and Reserve Design	Mycorrhizal Symbiosis	Ocean Acidification
Scale	O	U	C	U	C
Hierarchy	C	O	U	O	O
Cross-scale Interactions	O	C	O	O	C
Aggregation	R	R	R	R	R
Emergent Properties	R	C	O	O	C
Integration	O	O	R	C	R
Stability	C	C	C	R	O
Ecological Incorporation	O	O	O	O	O
Bottom-up/ Top-down	O	R	O	O	C
Resistance	C	C	C	O	C
Resilience	C	C	C	O	C
Importance of Scale Widely Recognized and Used in Analysis	C	U	U	C	C
Hierarchy Used as a Framework	R	C	C	R	R
Integration Across Scales	O	O	R	O	R

the gaps in our understanding of the scale and hierarchy of the mycorrhizal symbiosis.

As a proxy for larger field-based experimentation, mathematical models can serve as a mechanism by which we can integrate our understanding of the mycorrhizal symbiosis across scales. Johnson et al. (2006) identifies seven models that vary in their scale of ecological response, from individual to ecosystem-wide responses: functional equilibrium, economic, integrative agent-based, community feedback, coevolutionary mosaic, trophic food webs, and pedogenesis. Take for example the pedogenesis model. Mycorrhizal fungi are important to the formation of soil (pedogenesis) in that they create conditions that support the formation and stabilization of soil aggregates. ¬Physical variation in the mycorrhizal organ has been shown to alter soil aggregation development (Miller & Jastrow 1990). Research on small-scale processes within the mycorrhizal symbiosis has given us pathways to model and scale-up these effects. The pedogenesis model attempts to predict how microscopic interactions between mycorrhizae and their soil environment may lead to drastic bottom-up effects, impacting entire soil food webs. Although empirical studies alone usually lack integration across scales, data from these studies have been used to shape these models, allowing us to theorize how small-scale interactions can affect ecosystem processes.

SCALE AND HIERARCHY IN THE ECOLOGY OF TERRESTRIAL DISTURBANCE—Although imagery of terrestrial disturbances, such as treefalls and wildfires, often communicate devastation and death, such disturbances are crucial ecological events, creating heterogeneous landscapes that ultimately promote biodiversity. The success and survival of a vast array of species are dependent on a landscape that is dynamic and subject to change. Decades of research continually reinforced the idea that disturbance is necessary in order to create a mosaic of habitat types and stand ages that increases species diversity. However, there is a paucity of research that investigates the effects of disturbance across scales and hierarchical levels. Small-scale disturbances, such as treefalls, have far different implications than region-wide disturbances like wildfires. Furthermore, global disturbance events such as climate change are likely to have even more varied implications than small- and regional-scale disturbance. The effect of disturbance on an individual, a population, or species interactions highlights the need for the study of terrestrial disturbance within the context of hierarchy.

Pickett et al. (1989) attempted a system of concepts aimed to help ecologists deal with hierarchies in disturbance. Their system relies on a "minimal structure," consisting of the ecological entity of focus and its interacting components that allow this entity to persist. The authors provide an example of two ecologists wanting to study a population of southern pine: one studies productivity and the other pine beetle outbreak. The ecologist studying productivity would create a minimal structure model that includes energy fluxes as interactions and trophic levels as entities. The one studying pine beetle outbreak would model their minimal structure with canopy characteristics and soil resources as the entities, and phloem transport (which connects soil to canopy) as the interaction that connects the two, creating a persistent structure; the bark beetle population would then be external to this minimal structure and disturb the minimal structure. The development of this minimal structure model is exemplary of the importance of understanding hierarchy in disturbance ecology and provides ways in which we can integrate across scales.

Multiple studies in terrestrial disturbance ecology demonstrate that disturbance effects can vary depending on the scale of observation as well. Chaneton and Facelli (1991) evaluated the effect of disturbance on plant community diversity along a 5-m transect and a 1-ha (100 x 100 m) area. They found that comparisons among grassland conditions appeared scale-dependent. This may parallel meaningful changes in the relative importance of factors controlling species coexistence and community organization. In another multiscale study, Reed et al. (1993) examined the effects of spatial scale on the relationship between vegetation composition and underlying environmental variables. They found that as scale increased, so did the correlation with the physical environment, and confirmed that the results of vegetation analyses can depend greatly on the grain and extent of the samples employed.

Kotliar and Wiens (1990) explored how each level of hierarchical patch structure was influenced by the contrast among patches as well as the degree of aggregation of patches at lower levels in the hierarchy. The results of the study have wider implications in the study of habitat selection, population dynamics, and habitat fragmentation, and look to expand the realm of landscape ecology beyond the current focus on anthropocentric scales. Whittaker, Willis, and Field (2001) discuss the implications of considering multiple scale effects on the conclusions of narrowly focused studies. They state that one scale is not independent of other scales—a study narrowly focused on one scale may entirely miss the mechanism for what is observed. Additional studies that use scale and hierarchy as a framework continue to reinforce the idea that scale and hierarchy are essential when designing a study and interpreting its results, as one scale or level of hierarchy is not independent of those above and below the scale or level of hierarchy containing the study.

In the future it is essential to integrate scale and hierarchy into more studies involving terrestrial disturbance. This will be a daunting task to complete that will require more time and additional small-scale case studies with which to test generalizations among ecological systems. Integrating these findings into comprehensive meta-analyses across multiple scales and hierarchical levels can make more definitive conclusions about their effects and implications on ecological systems, as well as predictions for how these effects may vary in the future. This could greatly help conservation of biodiversity and mitigating economic and social losses when inevitable disturbance events occur.

OCEAN ACIDIFICATION: IMPLICATIONS OF SCALE AND HIERARCHY—Ocean acidification (OA) is a hot topic in marine research, as human activity continues to produce exorbitant amounts of atmospheric carbon dioxide (CO_2), which is absorbed by the ocean and ultimately increases acidity of oceans worldwide. This phenomenon affects the early life history of many marine organisms and even influences predator-prey dynamics (Dupont et al. 2008; Ferrari et al. 2011). Such observations of the impact of OA may reveal a cascade of effects on key trophic systems and other important mechanisms in ecology, but requires marine ecologists to consider these impacts in the frame of scale and hierarchy.

Scale and hierarchy are inherent in modern marine ecology due to the systematic interactions between species as well as spatial and temporal variation in marine systems. Marine ecosystems are being influenced globally, and with increased intensity, by large-scale top-down drivers such as OA. Key climatic processes occur both within short-term and long-term time scales. Over time, climatic effects that drive OA span globally and can carry effects from a single population up to the entire biosphere. A changing climate and transitions in localized weather patterns can lead to shifts in species distributions, disruption of match-mismatch systems, changes in migratory patterns, and other, largely undiscovered, effects.

Contemporary marine research attempts to quantify the effects of OA on marine ecosystems in the present, as well as the future, since the effects of OA vary with the time scale of study. The effects of global warming, OA, and their interactions are difficult to detect in short-term studies, but may manifest over time through changes in growth and behavior of organisms. These delayed effects may in turn affect ecosystem processes and structure (Godbold & Solan 2013). Feely et al. (2009) examined the uptake of anthropogenic CO_2 by the global ocean and its projected effects on seawater chemistry by the year 2100. They suggest that oceanic saturation of aragonite and calcite, which are crucial carbonate minerals, will be greatly reduced from current conditions. This will likely have detrimental effects on the ability of shell-building organisms and other marine calcifiers to sequester different forms of calcium carbonate.

Underlying hierarchical structure of marine systems has been examined in context of top-down effects triggered by OA. Shifts in the trophic structure of marine communities and changes in species dominances may occur, which could lead to the simplification of local food webs (Kroeker et al. 2011). Ocean acidification has also been shown to decrease diversity, biomass, and trophic complexity of marine communities, suggesting that biodiversity and ecosystem function are likely to suffer as the effects of OA intensify (Kroeker et al. 2011). In contrast, bottom-up effects of OA are commonly considered a challenge to study, as responses at large-scales are often caused by events occurring at much smaller scales on individual organisms. However, we argue that large-scale effects are often not limited to individuals, as is denoted in FIG 2; events at higher-order hierarchical levels may also cause repercussions on global scales.

Given the perceived limitations of cross-scale integration, attempts to predict the impacts of OA have been relatively limited to local or global scales, despite community

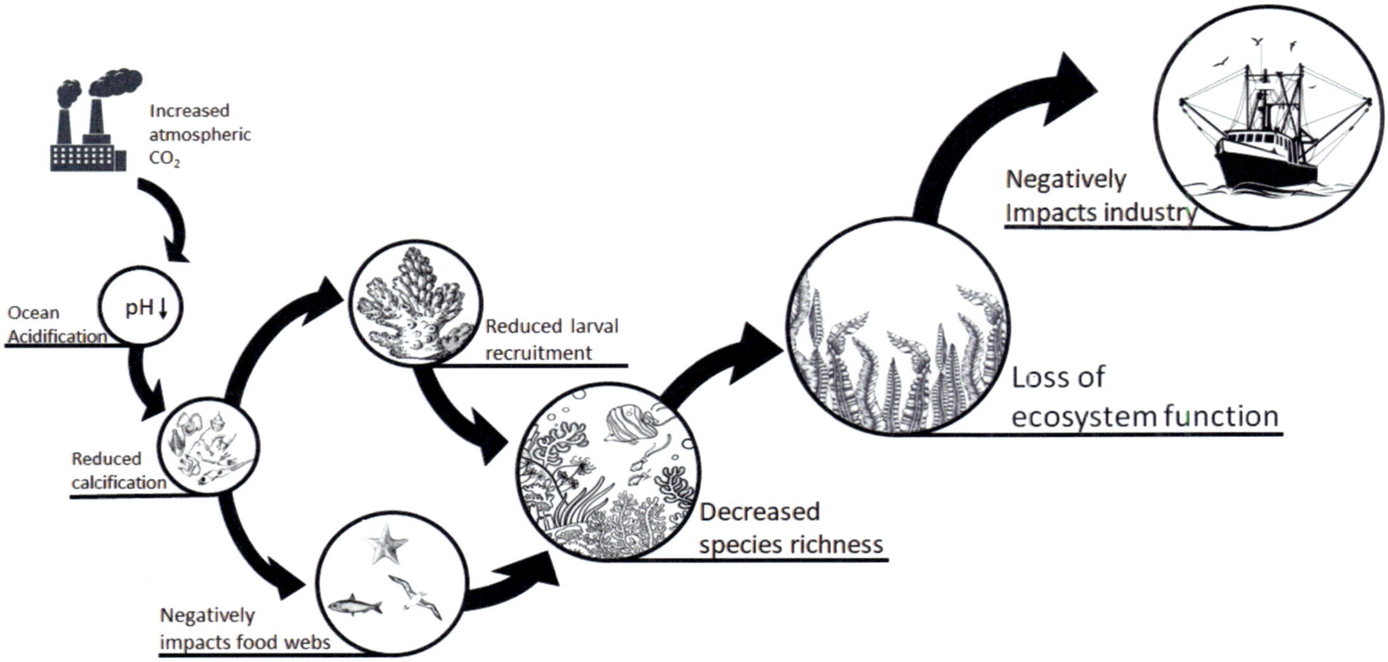

FIGURE 2. Ocean acidification as a conceptual framework.

or mesoscale studies being more meaningful for ecosystem management and resource use. Models that integrate sampling methods used in both larger- and smaller-scale analyses can help fill the sampling gap at intermediary scales. For example, Van Gennip et al. (2017) employed assimilative models to simulate and forecast animal movement patterns at a regional scale. If coupled with field experiments, investigators may predict community changes caused by stressors from large-scale sources such as OA.

THINGS CHANGE: HIERARCHY AND SCALE IN EVOLUTIONARY BIOLOGY—Evolution is incredibly complex, occurring at many different spatial scales and across levels of biological organization. Evolutionary processes also operate at multiple temporal scales as well, from the long term changes in traits to the tempo and mode of developmental processes. An understanding of these processes can inform us on changes in organisms through time. Evolution is often measured at a specific scale or organizational level. For example, changes in genetic structure are usually studied at the molecular and phenotypic levels, while phenotypic variation is studied at the population level and compared geographically at different temporal or spatial scales. However, the use of observations at different scales or biological levels can be insightful. This is especially true when describing new species, where both morphological and molecular evidence is used to support differences among populations.

Small-scale changes in an organism's genome can lead to drastic effects across hierarchical levels, affecting individuals and entire populations in positive, as well as negative, ways. Geneticist Dmitry K. Belyaev best described this in his seminal research on the genetic basis of animal domestication. Belyaev found that when select genes in an animal are slightly altered, they can give rise to a wide network of changes in the developmental processes the gene governs (Belyaev 1969). This "small" change at the genetic level led to profound changes at the phenotypic level, such as how tame an animal was likely to be (Belyaev 1969). Genotypic changes during domestication were even similar among mammals from different taxonomic groups, because they all shared similar regulatory mechanisms for hormones and neurochemistry. As a result, many phenotypes are shared among domesticated mammals, such as dwarf and giant varieties, piebald coat color, floppy ears, and changes in reproductive cycles (Belyaev 1969). Belyaev's research was, and still is, an incredible body of work, because it linked small changes at the genetic level to dramatic phenotypic and behavioral changes associated with domestication.

Unfortunately, not all genotypic changes lead to desirable phenotypic displays. The loss of genetic diversity and in-breeding often characteristic of endangered animal populations have led to pronounced physical deformities. In endangered Florida Panthers (*Puma concolor coryi*), individuals have been found with bent tails (Roelke et al. 1993). On the other hand, populations of invasive American Bull Frog (*Lithobates catesbeianus*) appear to not be negatively affected by a lack of genetic diversity and inbreeding (Klamath et al. 2016). These studies indicate that our current understanding of the emergent properties of low genetic diversity on individual fitness and population dynamics requires continued research and inquiry.

In addition to the hierarchical framework through which genetic selection operates, selection on life-history traits over ecological timescales can have far-reaching, indirect macroevolutionary effects. A shift in the relative timing between two developmental processes in a descendant ontogeny, also known as heterochrony, can often affect evolutionary processes (Raff & Wray 1989). A good example of heterochrony is displayed in obligate paedomorphic salamanders, which retain larval characteristics into adulthood. Because of this, populations of these salamanders are genetically more distinct from one another than are metamorphosing populations, which fully transform to a terrestrial adult stage and retain no larval characteristics in adulthood (Shaffer 1984).

High genetic diversity among populations of obligate paedomorphic salamanders is likely the result of low genetic flow, given that they tend to stay in smaller areas of streams or in isolated ponds. They may then have higher speciation rates compared to salamanders that can metamorphose and travel between ponds. Changes from facultative paedomorphosis to obligate paedomorphosis can occur in descendant ontogeny; therefore, paedomorphosis in salamanders can often result in heterochrony (Gould 1977). Interestingly, heterochrony at one hierarchical scale does not necessarily lead to heterochrony at other hierarchical scales. For example, the molecular basis of paedomorphosis in salamanders can vary among individuals and is not always related to shifts in developmental rates. In some salamanders, paedomorphosis is triggered by the disabling of the production or reception of a specific hormone, which is more related to turning a gene on or off rather than changing the timing of hormone production. While macromorphologically we see evidence of heterochrony in the form of delayed adult features, it was not triggered by a delayed molecular event. Therefore, changes in timing at the molecular or cellular level need not produce heterochronic patterns at the whole-organism level, and heterochrony at the organismal level need not involve changes in the timing of molecular events.

Evolutionary biologists frequently use hierarchy and scale as a framework, especially when understanding how changes at the genetic level can have long-term effects on individuals, populations, and even the persistence and conservation of entire species. Heterochrony and domestication are just a few examples that have employed integration across scales and hierarchies to better understand and explain complex evolutionary processes. With the advent of more robust genetic and molecular tools, we may uncover new information to help us integrate more intensively across hierarchical levels, therefore demystifying the complex and often enigmatic processes that collectively lead to evolution. As we enter the Earth's sixth mass extinction, the goal for future integration is to question how these effects spanning hierarchical levels and spatial scales interact and become linked or decoupled on ecological and evolutionary timescales.

USE OF SCALE AND HIERARCHY CONCEPTS IN RESERVE DESIGN RESEARCH—As human population continues to grow, ecological reserves play a major role in limiting human encroachment into important natural habitats forf sensitive wildlife. In many parts of the world, reserves represent the only places that natural biodiversity is maintained and natural processes are allowed to play out. Reserves must be designed for maximum conservation efficacy if they are to facilitate maintaince of biodiversity. Reserves that are too small and isolated can function essentially as ecological islands. Just like island systems, they may contain lower levels of biodiversity and be more prone to local extinctions (MacArthur & Wilson 1967). The use of scale and hierarchy as a framework for assessing reserve adequacy and efficacy has the potential to inform conservation strategies and policy decisions relating to land-use allocation.

Recognizing the concept of scale-dependency may play a key role in creating effective reserve systems, but logistics and limitations often prevent the integration of this concept into reserve design. Creating a small reserve in a single location may be helpful to species that have limited ranges that are contained within the reserve. However, highly mobile species with ranges that expand beyond the reserve's borders, such as migratory birds, may see little benefit if a network across the entirety of their range is not adequately preserved. Similarly, relatively sedentary species within a small reserve may be well protected on an individual or population level, but their genetic diversity may be severely limited if populations outside of the reserve go extinct or have no contact with the protected population.

Reserve size is a major theme in conservation biology. Baker (1992) noted that a key element of any reserve design is ensuring that reserves are large enough to withstand potential disturbance events. Pickett and Thompson (1978) also argued that reserves should be designed based on a "minimum dynamic area," a size threshold at which there can be enough internal repopulation within a reserve to avoid extinction after a disturbance. By considering scale-dependent effects when designing individual reserves and larger reserve systems, the preservation of biological diversity across all hierarchical levels could be increased significantly. Little work has been done showing how reserve design practices can actually be integrated across scales. Most studies recommend looking at larger scales for more mobile species and smaller scales for more sedentary ones. The few studies that have made recommendations on how to integrate across spatial scales when considering reserve designs cite a lack of comprehensive data as a major limiting factor (Andelman & Willig 2002).

There is clear evidence that scale can have significant implications on reserve design plans. Huber et al. (2010) compared algorithmically-generated reserve plans that were based on either local-level (within an individual county) or regional-level (within several neighboring counties) species abundance and distribution data and found the resulting reserves to have little overlap. Local planning seemed to ignore most large-scale ecological processes, but purely regional planning ignored resource specificity and habitat heterogeneity. Hartley and Kunin (2003) highlighted similar issues, noting how extinction risk and other conservation priority factors varied greatly depending on the scale at which they were examined. Both studies make similar recommendations that scale must be considered when making conservation plans and that data from multiple scales should be combined.

In addition to recognizing the importance of scale, some work has focused on choosing the appropriate scale when designing reserves. Schwartz (1999) compared the efficacy of both small- and large-scale reserve designs and recommended that if a single scale is to be considered, larger scales are generally preferred. When it comes to actually integrating multiple scales in a single reserve design, most studies recommend an algorithmic, computer modeling approach (Schwartz 1999; Huber et al. 2010;). But not many studies have actually put this approach into use. One study that did use this methodology while making recommendations for reserves to protect Paraguayan bats noted that the major limitation to this method is the lack of large, consistent data sets (Andelman & Willig 2002). While they found the results of the approach promising, they noted that it was only possible because they had access to long-term, comprehensive monitoring data for the bats in question.

If cross-scale integration is to become more common in the field of reserve design, large-scale and consistent data collection should be prioritized in areas of conservation concern. Such datasets would increase the ability of researchers to use computer modeling techniques to design more effective ecological reserves. Until these types of data become more readily available, scientists should at least acknowledge scale in their reserve plans and try to work at the scale most appropriate to their specific conservation goals. While integrating across spatial-scales may not yet be realistic, given the data limitations, the scale-dependence of conservation actions must be recognized when planning new conservation reserves. Reserve plans should focus on creating large enough reserves to maintain sufficient biodiversity, and networks of closely-linked reserves should be created if possible.

CONCLUSIONS—After reviewing the literature in each of the respective ecological disciplines, we developed a simple characterization of three different approaches or applications of scale and hierarchy in each ecological discipline. First, there are those applications that explicitly acknowledge the importance of scale or hierarchy in their discipline. There appears to be a growing number of applications that explicitly incorporate observations at two or more scales. Secondly, there are those papers that apply scale or hierarchy as a framework to help communicate how interactions across scales may be manifested at different hierarchical levels. Finally, there were those studies that integrated across scales or levels of hierarchy in a predictive capacity. Applications that integrate across scales or levels appear to be relatively rare and may still be primarily limited to theoretical studies or modeling exercises. The application of concepts related to scale and hierarchy in empirical studies has lagged compared to the rapid theoretical advances in the 1980s and 1990s. We hope that this paper inspires more thought on the importance of considering scale and hierarchy in ecology, its value as a framework, and the appeal of understanding the integrating ecological processes across scales and hierarchies into a few measurable and meaningful variables.

LITERATURE CITED

Allen, T. F. H., Hoekstra, T. W. 1992. Toward a unified ecology. New York: Columbia University.

Andelman, S. J., Willig, M. R. 2002. Alternative Configurations of Conservation Reserves for Paraguayan Bats: Considerations of Spatial Scale. *Conservation Biology* 16(5): 1352–1363. https://doi.org/10.1046/j.1523-1739.2002.01119.x.

Baker, W. L. 1992. The landscape ecology of large disturbances in the design and management of nature reserves. *Landscape Ecology* 7: 181–194.

Belyaev, D. 1969. Domestication of animals. *Science* 5: 47–52.

Chaneton, E. J., Facelli, J. M. 1991. Disturbance effects on plant community diversity: spatial scales and dominance hierarchies. *Vegetatio* 93(2): 143–155.

Dupont, S., Havenhand, J. H., Thorndyke, W., Peck, L. S., Thorndyke, M. 2008. Near-future level of CO_2-driven ocean acidification radically affects larval survival and development in the brittlestar Ophiothrix fragilis. *MEPS* 373: 285–294.

Feely, R. A., Sabine, C. L., Lee, K., Berelson, W., Kleypas, J., Fabry, V. J., Millero, F. J. 2004. Impact of anthropogenic CO_2 on the $CaCO_3$ system in the oceans. *Science* 305(5682): 362–366.

Fellbaum, C. R., Mensah, J. A., Cloos, A. J., Strahan, G. E., Pfeffer, P. E., Kiers, E. T., Bucking, H. 2014. Fungal nutrient allocation in common mycorrhizal networks is regulated by the carbon source strength of individual host plants. *New Phytologist* 203: 646–656.

Ferrari, M. C. O., McCormick, M. I., Munday, P. L., Meekan, M. G., Dixson, D. L., Lönnstedt, O., Chivers, D. P. 2012. Effects of ocean acidification on visual risk assessment in coral reef fishes. *Functional Ecology* 26(3): 553–558.

Fogel, R. 1980. Mycorrhizae and nutrient cycling in natural forest ecosystems. New Phytologist 86: 119–212.

Godbold, J. A., Solan, M. 2013. Long-term effects of warming and ocean acidification are modified by seasonal variation in species responses and environmental conditions. *Philos Trans R Soc Lond B Biol Sci* 368: 20130186. https://doi.org/10.1098/rstb.2013.0186.

Gould, S. J. 1977. *Ontogeny and phylogeny*. Harvard University Press, Cambridge.

Hartley, S., Kunin, W. E. 2003. Scale dependency of rarity, extinction risk, and conservation priority. *Conservation Biology* 17(6): 1559–1570. https://doi.org/10.1111/j.1523-1739.2003.00015.x.

Huber, P. R., Greco, S. E., Thorne, J. H. 2010. Spatial scale effects on conservation network design: trade-offs and omissions in regional versus local scale planning. *Landscape Ecology* 25(5): 683–695. https://doi.org/10.1007/s10980-010-9447-4.

Johnson, N. C., Hoeksema, J. D., Bever, J. D., Chaudhary, V. B., Gehring, C., Klironomos, J., Koide, R., Miller, R. M., Moore, J., Moutoglis, P., Schwartz, M., Simard, S., Swenson, W., Umbanhowar, J., Wilson,

G., Zabinski, C. 2006. From Lilliput to Brobdingnag: Extending models of mycorrhizal function across scales. *Bioscience* 56(11): 889–900.

Kotliar, N. B., Wiens, J. A. 1990. Multiple scales of patchiness and patch structure: A hierarchical framework for the study of heterogeneity. *Oikos* 59(2): 253–260.

Kroeker, K. J., Micheli, F., Gambi, M. C., Martz, T. R. 2011. Divergent ecosystem responses within a benthic marine community to ocean acidification. *Proceedings of the National Academy of Sciences of the United States of America* 108(35): 14515–14520.

Levin, S. A. 1992. The problem of pattern and scale in ecology. *Ecology* 73: 1943–1967.

MacArthur, R. H., Wilson, E. O. 1967. *The theory of island biogeography*. Princeton University Press, Princeton, New Jersey.

Miller, R. M., Jastrow, J. D. 1990. Hierarchy of root and mycorrhizal fungal interactions with soil aggregation. *Soil Biology and Biochemistry* 22(5): 579–584.

Morris, J. L., Puttick, M. N., Clark, J. W., Edwards, D., Kenrick, P., Pressel, S., Wellman, C. H., Yang, Z., Schneider, H., Donoghue, P. C. J. 2018. The timescale of early land plant evolution. *Proceedings of the National Academy of Sciences of the United States of America* 115(10): E2274–E2283.

O'Neil, R. V. O, King, A. W. 1996. Homage to St. Michael; or why are there so many books on scale. In D.L. Peterson and V. T. Parker, Eds. *Ecological Scale: Theory and Applications*. Columbia University Press, New York.

Pickett, S. T. A., Thompson, J. N. 1978. Patch dynamics and the design of nature reserves. *Biological Conservation* 13(1): 27–37.

Pickett, S. T. A., Kolasa, J., Armesto, J. J., Collins, S. L. 1989. The ecological concept of disturbance and its expression at various hierarchical levels. *Oikos* 54: 129–136.

Raff, R. A., Wray, G. A. 1989. Heterochrony: developmental mechanisms and evolutionary results. *Journal of Evolutionary Biology* 2: 409–434.

Reed, R. A., Peet, R. K., Palmer, M. W., White, P. S. 1993. Scale dependence of vegetation-environment correlations: A case study of a North Carolina piedmont woodland. *Journal of Vegetation Science* 4: 329–340.

Roelke, M. E., Martenson, J. S., O'Brien, S. J. 1993. The consequences of demographic reduction and genetic depletion in the endangered Florida panther. *Current Biology* 3(6): 340–350.

Schwartz, M. W. 1999. Choosing the appropriate scale of reserves for conservation. *Annual Review of Ecology and Systematics* 30(1), 83–108. https://doi.org/10.1146/annurev.ecolsys.30.1.83.

Shaffer, H. B. 1984. Evolution in a paedomorphic lineage. I. An electrophoretic analysis of the Mexican ambystomatid salamanders. *Evolution* 38: 1194–1206.

Simard, S. W., Perry, D. A., Jones, M. D., Myrold, D. D., Durall, D. M., Molina, R. 1997. Net transfer of carbon between ectomycorrhizal tree species in the field. *Nature* 338: 579–582.

Smith, S. E., Read, D. J. 1997. *Mycorrhizal symbiosis*. 2nd edn. Academic, San Diego.

van Gennip, S. J., Popova, E. E., Yool, A., Pecl, G. T., Hobday, A. J., Sorte, C. J. B. 2017. Going with the flow: The role of ocean circulation in global marine ecosystems under a changing climate. *Global Change Biology* 23(7): 2602–2617.

Wang, B., Qiu, Y. L. 2006. Phylogenetic distribution and evolution of mycorrhizas in land plants. *Mycorrhiza* 15(5): 229–363.

Whittaker, R. J., Willis, K. J., Field, R. 2001. Scale and species richness: Towards a general, hierarchical theory of species diversity. *Journal of Biogeography* 28(4): 453–470.

SUPPLEMENTARY MATERIAL

TABLE S1. List of terms and concepts related to scale and hierarchy in ecology..

aggregation	Grouping of organisms or biological processes that may affect the quality of information one can extrapolate depending on the scale of the observations.
bottom-up/to-down	Ecosystem structuring driven by nutrient supply, productivity, and type of primary producers (bottom-up) or top predators controlling the structure or population dynamic (top-down).
cross-scale interations	Relationship of organisms or processes across a broad range of space or time.
ecological incorporation	The ability of an ecosystem to adapt to changes.
emergent properties	A property of a system that is not seen at the level of the individuals or characteristics that comprise that system.
hierarchy	The arrangement and relation of organisms to each other.
integration	How various levels of hierarchy relate to and compose other levels.
resilience	The ability to resist damage and recover quickly after disturbance.
resistance	The ability to remain fundamentally unchanged by disturbance.
scale	Spatial and temporal size.
stability	The ability of an ecosystem to return to equilibrium after a disturbance.

Evaluating the Effect of Time of Day on Singing Behavior in Anna's Hummingbirds

Adrian D. Macedo[1*], Maxine R. Mota[2]

KEYWORDS—*behavior, birds calls,* Calypte anna, *hummingbirds, Pacific Northwest, bird song*

INTRODUCTION—Sexual selection has driven the evolution of numerous strategies to attract mates. In songbirds, the most recognized method is the intricate behaviors of song production. However, there are other factors that also influence song behavior in birds. It has long been observed that songbirds put forth the most singing effort during dawn and dusk. Reasons for this temporal pattern in song have raised many hypotheses, such as singing at different times of day may possibly minimize interspecific acoustic interference (Cody and Brown 1969). Alternatively, songs are carried farther and clearer during the dawn and dusk, termed the acoustic transmission hypothesis (Henwood and Fabrick 1979). Empirical support for these hypotheses is equivocal, as they have been supported in some studies but not in others (Keast 1994; Catchpole and Slater 1995; Staicer et al. 1996; Brown and Handford 2002; Dabelsteen and Mathevon 2002).

We wanted to determine if the dawn/dusk pattern is used by non-passerines such as hummingbirds. This temporal pattern in song has been studied in the lekking behavior of white-bellied emeralds (*Amazilia candida*) and Swallow-tailed hummingbirds (*Eupetomena macroura*) (Atwood et al. 1991; Pizo and Silva 2001). Lekking is when the males of one species get together in one area to advertise for mates. Although they do not exhibit lekking, Anna's hummingbirds (*Calypte anna*) employ song, dipping, and shuttle display behaviors in their mating rituals. The effect of time of day on the singing behavior of Anna's hummingbirds has not been described in detail. Perhaps Anna's hummingbirds also sing more frequently at dawn and dusk like the swallow-tailed hummingbirds, white-bellied emeralds, and numerous other birds, or perhaps there is no temporal effect on their singing behavior.

*Corresponding Author: adm566@humboldt.edu
[1]Graduate Student
Department of Biological Sciences
Humboldt State University, Arcata, CA 95521
[2]Administrative Support Assistant
Learning Center
Humboldt State University, Arcata, CA 95521

Macedo, A. D., Mota, M. R. 2019. Evaluating the effect of time of day on singing behavior in Anna's hummingbirds. *ideaFest Journal: Interdisciplinary Journal of Creative Works & Research from Humboldt State University* 3: 24–28.

The Anna's hummingbird is a 3–6 g, medium-sized hummingbird with a straight shortish bill and a broad tail (Russell 1996). When perched, their wingtips meet the tip of their short tails (Russell 1996). Their color pattern is composed mostly of green and gray plumage, without any rufous or orange marks on the body (Russell 1996). Males have iridescent reddish-pink feathers covering their head and throat; this area is referred to as a gorget (Russell 1996).

In the first half of the 20th century, the Anna's hummingbird bred only in northern Baja California and southern California. The planting of exotic flowering trees in gardens and other human-populated areas provided food and nesting resources that have allowed Anna's hummingbirds to expand their range north, as far as British Columbia (Russell 1996). Their primary food source is nectar from flowering plants including currants and gooseberries (*Ribes* spp.), manzanita (*Arctostaphylos* spp.), introduced eucalyptus (*Eucalyptus* spp.), and many others (Russell 1996). They also eat a wide array of smaller insects (Russell 1996).

Behavioral aspects of the male Anna's hummingbird include territoriality, in which resident males will defend a particular area throughout the year (Woods 1940;

Pitelka 1942). They defend this area to protect food and breeding resources (Ewald and Carpenter 1978; Ewald and Bransfield 1987). Behaviors often associated with territorial defense include chasing, display dives, aggressive vocalization, and chatter-sway. The display dive involves a male hummingbird flying upwards of 740 m then diving toward the ground making a sharp chirp sound by pushing wind through its tail feathers (Bent 1940; Banks and Johnson 1961; Clark and Feo 2008). The chatter-sway occurs before chasing an intruder, by vocalizing and swaying back and forth while perched to warn intruders (Stiles 1982).

The behaviors associated with mate attraction include those used in territorial defense (chasing, display dives, and song vocalizations), but shuttle displays are used only in mate attraction. The display dive is generally an aggressive behavior, however is sometimes used to attract a mate (Stiles 1982). Song vocalization to attract mates is distinctly different than aggressive vocalizations (Stiles 1982). Shuttle displays are characterized by the male hovering in front of the female and flying rapidly back and forth in tight arcs while singing (Stiles 1982).

The aim of our study was to test the effect of time of day on the number of songs performed per unit of time (NSPT) in Anna's hummingbirds. We tested the hypothesis that time of day influences the NSPT of Anna's hummingbirds. We predicted that Anna's hummingbirds would follow a pattern of dawn/dusk singing similar to many passerines. If they do follow this pattern, then we would expect to find a higher frequency of song in the morning and evening hours of the day.

METHODS—*Study area*. We conducted our study on the Humboldt State University (HSU) campus and in the surrounding neighborhoods of Arcata, California, located in the Pacific Northwest region of the USA. Temperatures in Arcata average 7–15 °C over the year with an average rainfall of 101 cm per year. Our study was carried out from March to April, during the middle of the breeding season for the Anna's hummingbird (Woods 1940). The study area is in a semi-urban environment consisting mainly of human infrastructures such as large buildings, roads, and houses. However, there are many pockets of vegetation scattered throughout the area. The types of vegetation within this study area vary greatly as most of the plant communities are the result of intentional planting by humans for ornamental value and food production. Some of the more common plant families found in the study area include Rosaceae, Ericaceae, Pinaceae, and Liliaceae. Humboldt State functions as a college campus and has a high amount of human activity during the daytime hours when students and other community members make use of the resources available on campus. This also means that there is a great deal of pedestrian and motor traffic throughout the surrounding neighborhoods as people make their way to and from campus. The eastern border of the study area consisted of the Arcata Community Forest (ACF), a mixed-conifer forest dominated by coastal redwood (*Sequoia sempervirens*), and private property, which is of the same forest type as the ACF. We designated the other borders of the study area using K Street/Alliance Avenue as the western border of the study area while the southern border was Samoa Boulevard. The northern border was the Mad River (FIG 1).

Field observation. To gather data on the singing behavior of Anna's hummingbirds, we conducted focal animal observations of male Anna's hummingbirds during different times of the day in order to see if there was a difference in the number of songs performed at different hours. In order to locate male Anna's hummingbirds, we

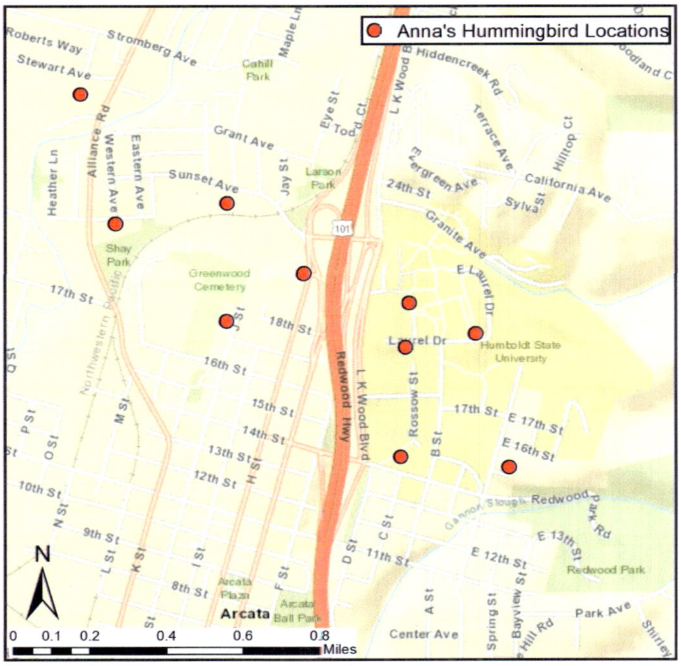

FIGURE 1. Individual male Anna's hummingbird locations in Arcata, northern California, USA.

conducted pre-study surveys of the study area looking for males and listening for the distinctive mate attraction vocalizations. At each place where a male hummingbird was positively identified, we marked and numbered 10 sites on a map for use during the study. Each site was at least 50 m from other sites, so it could be considered a separate individual male hummingbird as some male hummingbirds can have core territory areas up to 2,023 m^2 (Pitelka 1951).

When choosing a focal bird to observe for the day, we randomly chose a hummingbird site from among all sites using a random number generator. We then went to the site and attempted to locate the hummingbird within 15 min of arrival. If the bird was not located within 15 min, then we randomly chose another site for observation. If the bird was located within 15 min, we started a stopwatch within 1 min of locating the bird either by sight or sound. We continuously recorded each song vocalization event that happened within a 15-min time frame at the site at three different times throughout the day: morning (7am–9am), midday (12pm–2pm), and evening (5pm–7pm). The song vocalization events that happened during the observation period were then quadrupled to give us an estimate of the number of songs per hour, analogous to NSPT. A new site was randomly chosen each observation day until all hummingbird sites were observed at least once at each of the three chosen times.

We also recorded weather, temperature, time, and date during each observation. Weather was categorized as sunny, partly cloudy, overcast, or light rain. We did not conduct surveys during conditions considered heavy rain due to the lack of visibility. Additionally, we worked under the assumption that most hummingbirds would not be active during heavy rain due to the significant increases in the energy needed for flight during heavy rain (Ortega-Jimenez and Dudley 2012).

We analyzed the data by using time of day as an independent variable and song vocalization frequency as a dependent variable. We completed one survey of each individual at each of the time categories. The NSPT for morning, midday, and evening had a normal distribution and equal variances according to Levene's test and Shapiro-Wilk normality test, therefore an ANOVA test was used.

RESULTS—A total of 10 individual male Anna's hummingbirds were observed in their territories for a total of 150 min for all time groups from 1 Mar 2017 through 5 Apr 2017. Average NSPT was greater in the morning and the evening compared to average NSPT at midday, however these differences were not significant (FIG 2). Time of day did not have a significant effect on the pattern of NSPT within our sample population ($F = 0.355$, df = 2, $P = 0.704$).

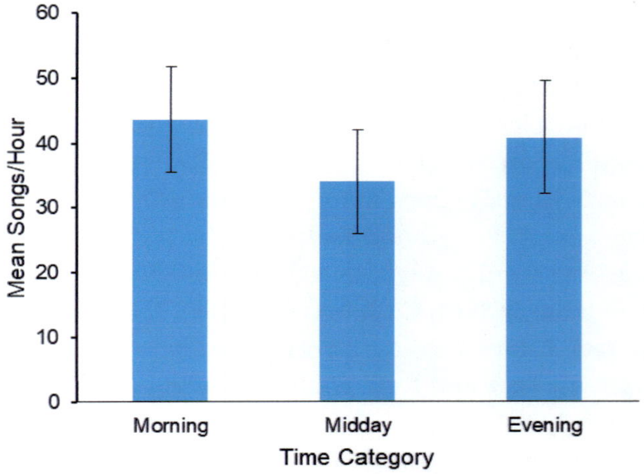

FIGURE 2. Overall mean number of songs per hour (± SE) of Anna's hummingbirds (n = 10) during different times of day (morning, midday, evening) in Arcata, CA, USA, 19 Apr 2016.

DISCUSSION—We surveyed male Anna's hummingbirds throughout a semi-urban area at different times of day to see the effect that time of day had on NSPT. Overall, our results do not support our hypothesis that Anna's hummingbirds will have a significantly greater NSPT during the morning and evening than during midday. Despite observable differences in NSPT between the time categories, the differences were not statistically supported. These results may be due to the sample size, confounding variables, and variation in the breeding behavior among different male hummingbirds. Other possible variables that may have confounded our results was variation in breeding effort over the survey period, weather conditions, and temperature. There was observable variation among individual's NSPT, with some birds singing more than others (FIG 3). Some males sung much more frequently; for example, bird 9 had a maximum NSPT of 88 compared to bird 5 who's maximum NSPT was one. This variation in NSPT could be related to a difference in testosterone levels among males, causing some to sing more and others to sing less. This has been experimentally tested in Bengalese Finches. Males of this species who have higher levels of testosterone sang more than birds with less testosterone (Ritschard et al. 2011).

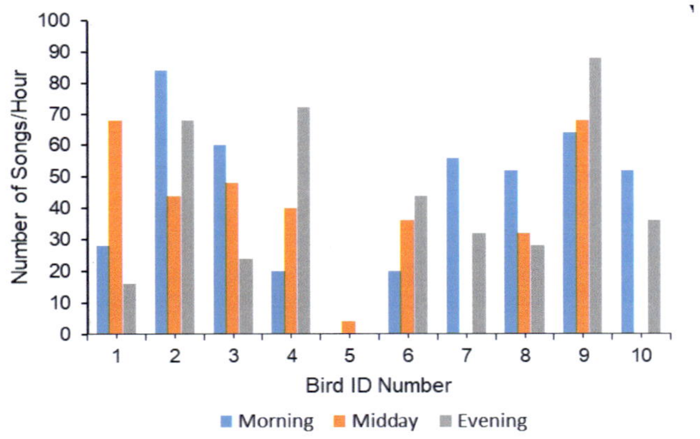

FIGURE 3. Number of songs per hour (± SE) of individual male Anna's hummingbirds (n = 10) during different times of day (morning, midday, evening) in Arcata, CA, USA, March–April 2016.

Survey date, specifically when the periods were split between the weeks of 10 Mar 2017 through 19 Mar 2017 and 26 Mar 2017 through 5 Apr 2017, experienced a significant decline in number of songs per hour between the two periods (t = 2.9738, df = 14.606, $P = 0.009687$) using a Welsh two-sample t-test (FIG 4). Therefore, this significant change in breeding effort is a valuable finding and may explain the minimized response of time of day on NSPT. There were no observable patterns in the NSPT of Anna's hummingbirds in different temperature and weather conditions. This was likely because there were more sampling periods in overcast weather than other conditions, and temperature stayed between 12 °C and 15 °C with less variation. Given a larger data set with more variation and an even distribution of weather observations and temperature, we would have more evidence to identify stronger trends with weather and NSPT (FIG 5). Temperature may also have influenced NSPT, but we were unable to analyze this trend as we did not record temperature for all our observations (FIG 6).

These results suggest that males reduced their territorial defense and attempts to attract mates over the course of the breeding season. This has not been studied much in hummingbirds, but our observations suggest that as it gets later in the season, there is a decrease in male song vocalizations and dive displays to attract mates. This has been the case in passerines such as Stonechats (*Saxicola torquata*), which stop singing completely once hatchlings have emerged and females are no longer receptive to breeding (Greig-Smith 1982).

As with all studies, a large sampling size and sample replication are often required to minimize the effects of biases and confounding variables. For greater statistical power we would need to increase our sample size to encompass a larger portion of the male Anna's hummingbirds that live within Arcata as well as a larger sample size of weather types, temperatures, and over a longer course of time.

Although our hypothesis was not significantly supported, there is a trend toward male Anna's hummingbirds having little daily temporal selection pressure—they sing throughout the day with no concentration of song behaviors during a certain period of the day. This could be due to a lack of acoustic interference and miss-transmission at the sites where the hummingbirds were observed. Further research is needed to confirm our results, which could then be used within bird conservation, especially with similar but more cryptic species such as South American hummingbirds.

MANAGEMENT IMPLICATIONS—Knowledge on the mating behaviors of a species is vital to wildlife

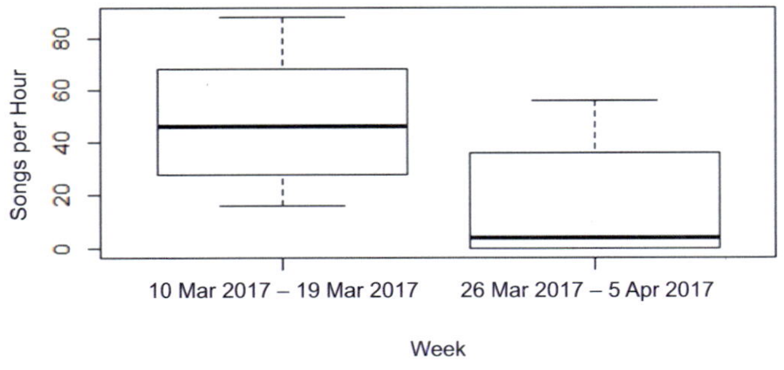

FIGURE 4. The mean number of songs per hour of Anna's hummingbirds (n = 10) during the data collection period in Arcata, CA, USA, March–April 2016.

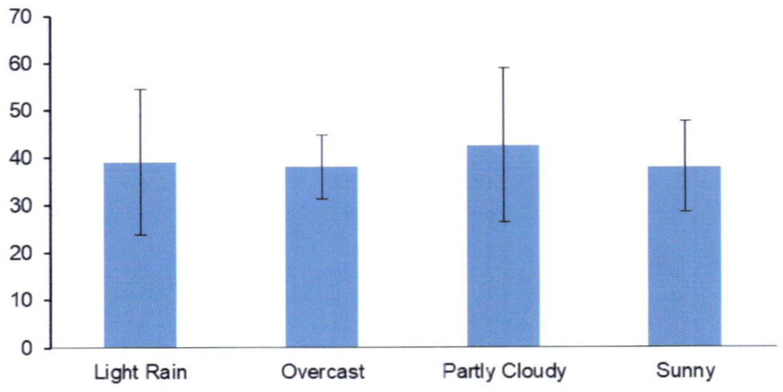

FIGURE 5. The mean number of songs per hour (± SE) of Anna's hummingbird (n = 10) during different weather conditions in Arcata, CA, USA, March–April 2016.

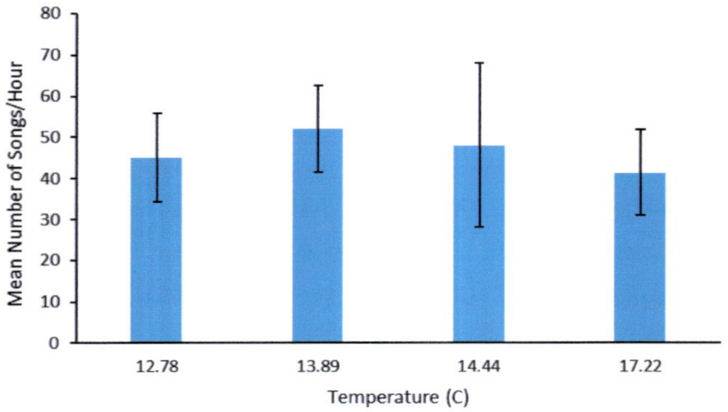

FIGURE 6. The mean number of songs per hour (± SE) of Anna's hummingbirds under the effect of varying temperatures in Arcata, CA, USA, March–April 2016.

managers when they are attempting to create appropriate species management plan. This is especially important in areas that have high amounts of human noise, which may disturb the breeding behavior of bird species that rely on acoustic transmission (Patricelli and Blickley 2006). Even though it was not statistically supported, there were differences in the frequency of song of Anna's hummingbirds thought the day, with more singing during the morning and evening than at midday. Additionally, the frequency of song overall significantly decreased across the survey period indicating that there is more breeding effort during early March compared with later in the season in Arcata, CA. This suggests that certain times of the year and day are used for completing most mating activities. While the Anna's hummingbird is not a threatened species, for a hummingbird species of concern, it may be important to minimize disturbance during peak breeding times to increase the chances of successful breeding. However, further studies must be done to confirm when most breeding activities occur for other hummingbirds as there could be variation within different species that would require different management strategies.

ACKNOWLEDGMENTS—We would like to thank Dr. Barbara Clucas and Dr. Mark Cowell for their advice and guidance throughout this study. Without their feedback and advice this study would not have been completed.

REFERENCES

Atwood, J. L., Fitz, V. L., Bamesberger, J. E. 1991. Temporal patterns of singing activity at leks of the white-bellied emerald. *Wilson Bull* 103: 373–386.

Banks, R. C., Johnson, N. K. 1961. A review of North American hybrid hummingbirds. *Condor* 63: 3–28.

Bent, A. C. 1940. Life histories of North American cuckoos, goatsuckers, hummingbirds, and their allies. *U. S. National Museum Bulletin*: 176.

Brown, T. J., Handford, P. 2002. Why birds sing at dawn: The role of consistent song transmission. *Ibis* 145: 120–129.

Catchpole, C. K., Slater, P. J. B. 1995. *Bird song: Biological themes and variations*. Cambridge: Cambridge University Press.

Clark, C. J., Feo, T. J. 2008. The Anna's hummingbird chirps with its tail: A new mechanism of sonation in birds. *Proceedings of the Royal Society B: Biological Sciences* 275: 955–962.

Cody, M. L., Brown, J. H. 1969. Song asynchrony in chapparal birds. *Nature* 272: 778–780.

Dabelsteen, T., Mathevon, N. 2002. Why do songbirds sing intensively at dawn? *Acta Ethologica* 4: 65–72.

Edwald, P. W., Bransfield, J. 1987. Territory quality and territorial behavior in two sympatric species of hummingbirds. *Behavioral Ecology and Sociobiology* 20: 285–293.

Ewald, P. W., Carpenter, F. L. 1978. Territorial responses to energy manipulations in the Anna's hummingbird. *Oecologia* 31: 277–292.

Greig-Smith, P. W. 1982. Seasonal patterns of song production by male Stonechats *Saxicola torquata*. *Ornis Scandinavica* 13: 225–231.

Henwood, K., Fabrick, A. 1979. A quantitative analysis of the dawn chorus: Temporal selection for communicatory optimization. *American Naturalist* 114: 260–274.

Keast, A. 1994. Temporal vocalization patterns in members of a eucalypt forest bird community: The effects of weather on song production. *Emu* 94: 172–180.

Ortega-Jimenez, V. M., Dudley, R. 2012. Flying in the rain: Hovering performance of Anna's hummingbirds under varied precipitation. Proceedings of the Royal Society of London B: Biological Sciences 279: 3996–4002.

Patricelli, G. L., Bickley, J. L. 2006. Avian communication in urban noise: Causes and consequences of vocal adjustment. *The Auk* 123: 638–649.

Pitelka, F. A. 1942. Territoriality and related problems in North American hummingbirds. *The Condor* 44: 189–204.

Pitelka, F. A. 1951. Ecologic overlap and interspecific strife in breeding population of Anna and Allen Hummingbirds. *Ecology* 32: 641–661.

Pizo, M. A., Silva, W. R. 2001. The dawn lek of the swallow-tailed hummingbird. *The Wilson Ornithological Society* 113: 388–397.

Ritschard, M., Laucht, S., Dale, J., Brumm, H. 2011. Enhanced testosterone levels affect singing motivation but not song structure and amplitude in Bengalese finches. *Physiology & Behavior* 102: 30–35.

Russell, S. M. 1996. Anna's hummingbird (*Calypte anna*). The Birds of North America, No. 226 (A. Poole and F. Gill, eds.). The Birds of North America Online, Ithaca, New York.

Staicer, C. A., Spector. D. A., Horn, A. G. 1996. The dawn chorus and other diel patterns in acoustic signaling. Cornell University Press: Ecology and Evolution of Acoustic Communication in Birds: 426–453.

Stiles, G. F. 1982. Aggressive and courtship displays of the male Anna's hummingbird. *The Condor* 84: 208–225.

Woods, R. S. 1940. *Calypte anna* Anna's Hummingbird, p. 371–387. In A. C. Bent, Life histories of North American cuckoos, goat- suckers, hummingbirds, and their allies. U.S. National Museum Bulletin: 176.

Responses of *Bouteloua eriopoda* and Soil Stability to Precipitation Extremes in Chihuahuan Desert Grassland

Laura K. Sadorf[1*]

ABSTRACT—Climate change is expected to increase the frequency and duration of extreme dry and wet years. As water is the most limiting resource in these semi-arid deserts, it is important to understand how semi-arid plants respond to precipitation changes, given that climate change will alter desert ecosystems in the future. To study these responses, I applied an extreme precipitation treatment (10 mm water twice per week) to a *Bouteloua eriopoda*-dominated desert grassland preceding the 2018 monsoon season. I measured soil moisture, soil stability, and vegetative growth of *B. eriopoda* within treatment and control plots, which received no water addition. While two natural rain events increased *B. eriopoda* growth in all plots, growth was significantly increased in response to the extreme precipitation treatment. Yet, control plots responded more quickly to natural rain pulses than continuously watered plots. Soil stability was not correlated with total cover of *B. eriopoda*. Soil stability was also measured at two adjacent long-term precipitation manipulation experiments. In comparison with this experiment, soils were more stable at watered plots in a long-term monsoon rainfall addition experiment and control plots in a long-term drought experiment. These results indicate that extreme precipitation events are beneficial for *B. eriopoda*-dominated grasslands, and *B. eriopoda* is especially responsive to rain pulses between dry periods. Further investigation is needed to determine the interaction between *B. eriopoda* and soil stability, which may provide insight for future success of this dominant desert grass.

KEYWORDS—*black grama,* Bouteloua eriopoda, *phenology, soil stability, precipitation, semi-arid, pulse-dynamics, productivity*

INTRODUCTION—Global climate models predict a more variable climate in the future, including increased frequency of extreme wet and dry years (Knapp et al. 2015). Currently, extreme wet years are rare, and their impacts should be documented through means other than modeling, such as experimental approaches (Knapp et al. 2015; Knapp et al. 2017). Many studies describe sporadic rainfall events via the pulse-dynamics paradigm (Ogle & Reynolds 2004; Reynolds et al. 2004; Thomey et al. 2011; Báez et al. 2013; Collins et al. 2014; Petrie et al. 2014; Thomey et al. 2014). Noy-Meir (1973) first described pulse-dynamics in arid land ecosystems as the biological processes that result after a rainfall event. This refers to sporadic rain events that drive biological responses of plants, such as growth and reproduction (Ogle & Reynolds 2004). More recently, the pulse-reserve paradigm is defined as rain pulses that drive biological activity resulting in a resource reserve, such as biomass or available nutrients (Collins et al. 2014). This reserve is important for driving plant responses well after a rain event (i.e. pulse) has occurred (Collins et al. 2014). According to Reynolds et al. (2004), a "pulse" is categorized as a collection of events that are capable of recharging soil water for a significant portion of the season. For example, plant productivity in semi-arid ecosystems responds rapidly to rainfall pulses, and thus is driven by pulse dynamics. Productivity during the growing season is therefore a function of the frequency and intensity of precipitation events (Reynolds et al. 2004; Collins et al. 2014), which are predicted to become more extreme under climate change (Knapp et al. 2015).

Many studies have looked at pulse-reserve dynamics

*Corresponding Author: lks147@humboldt.edu
[1]Undergraduate Student
Department of Biological Sciences
Humboldt State University, Arcata, CA 95521

Sadorf, L. K. 2019. Responses of *Bouteloua eriopoda* and soil stability to precipitation extremes in Chihuahuan desert grassland. *ideaFest Journal: Interdisciplinary Journal of Creative Works & Research from Humboldt State University* 3: 29–35.

across varying lengths and degrees of precipitation treatments (Thomey et al. 2011; Báez et al. 2013; Thomey et al. 2014). Báez et al. (2013) compared plant community dynamics of *Bouteloua eriopoda* and *Larrea tridentata* dominated sites after 4–5 years of experimental drought and precipitation. This long-term rainfall manipulation experiment showed that *B. eriopoda* is sensitive to drought—the response is similar to shallow-rooted grasses despite the fact that *B. eriopoda* is a C_4 plant, which typically thrives in hot, dry climates (Báez et al. 2013). Thomey et al. (2011) conducted a rainfall manipulation experiment and compared aboveground net primary productivity (ANPP) of *B. eriopoda* after small, large, and ambient rainfall treatments over two monsoon seasons. Large rainfall treatments were found to have a greater effect on plant growth than small rainfall treatments, suggesting that *B. eriopoda* maximizes carbon gain during advantageous periods (Thomey et al. 2011). Extreme precipitation experiments are important in *B. eriopoda*-dominated grasslands because this C_4 grass is sensitive to drought (Báez et al. 2013; Thomey et al. 2014), and water is the most limiting resource for plant productivity in all grass-dominated ecosystems (Wilcox et al. 2017).

Although the majority of extreme precipitation and drought studies measure ANPP as an indicator of grass production in response to rainfall treatments, plant vegetative growth can be a better indicator of plant growth on a short time scale, which can represent a more accurate depiction of when plants are fixing carbon (Wang et al. 2018). Precipitation pulses, and thus pulse-reserve dynamics, are biologically important because they can affect plant phenology by stimulating growth of vegetative and reproductive structures (Noy-Meir 1973; Ogle & Reynolds 2004). Ogle & Reynolds (2004) elaborated on pulse-reserve dynamics by noting that the response (i.e. growth rate) of a plant depends on its prior state before rainfall. The prior state refers to antecedent conditions, such as activity level of roots or enzymes, which influence the water uptake and photosynthetic rate of plants (Ogle & Reynolds 2004). This elaboration included thresholds that govern plant responses, such as the maximum growth rate a plant exhibits after precipitation pulses (Ogle & Reynolds 2004). An increase in the growth rate of a plant depends on the size of the pulse. If the growth rate was near the plant's maximum threshold before a pulse, then additional pulses would have no effect on its growth, as the plant's growth rate is already at a maximum. Vegetative growth is a good indicator of plant response to pulses, because it visually demonstrates the uptake of water and nutrients from the soil (Adair & Blake 2010; Wang et al. 2018). Even small rain events can stimulate resource pulses by supporting activity of biological soil crusts ("biocrusts") and only a slight increase in rain events can stimulate plant nutrient uptake and photosynthesis (Adair & Burke 2010).

According to Collins et al. (2008), there are important interactions and feedbacks of biotic and abiotic processes that assist in governing pulse-reserve dynamics. These biotic and abiotic processes include biocrust processes, soil moisture dynamics, and soil stability (Collins et al. 2008; Collins et al. 2014). Soil stability is defined as the degree to which soils resist erosion and is affected by biotic and abiotic processes, though biological components contribute the most to stability (Chaudhary et al. 2009; Fernandes et al. 2018). Soil stability tests are considered important for assessing ecosystem health according to Rangeland Monitoring Procedures (Eldridge & Greene 1994; Herrick et al. 2015). Biocrusts provide soil stabilization services through filamentous cyanobacteria located in the upper 2 mm soil layer (Ferrenberg et al. 2015; Fernandes et al. 2018). The link between soil stability and plant vegetative growth may be drawn from the ability of biocrusts to regulate soil hydrology and nutrient cycling through their rapid response to rain pulses (Ferrenberg et al. 2015).

This study was designed to elucidate possible connections between *B. eriopoda* vegetative growth and soil stability after a precipitation treatment to represent pulse-dynamics, via the experimental replication of summer monsoon events. The precipitation treatment was designed to represent summer storm events because precipitation in the Chihuahuan Desert is highly variable and summer monsoons produce 53% of the average annual precipitation (Reynolds et al. 2004; Thomey et al. 2011). This study set out to test the following questions: (i) what effects do extreme precipitation additions have on the vegetative growth of *B. eriopoda*; (ii) what are the effects of these same treatments on soil stability; and (iii) how does soil stability differ in response to a short-term monsoon in comparison to long-term monsoon and drought experiments?

Specifically, this study hypothesizes that: (i) extreme precipitation treatments will increase the total biomass of *B. eriopoda*, which will be evident in the presence of new foliage; (ii) increasing water to the system will increase soil stability; (iii) high soil stability will retain water longer, allotting more time for *B. eriopoda* roots to uptake water, thus causing an increase in foliar cover; and (iv) low soil stability will retain water in the soil for a shorter period of time, causing low water uptake in *B. eriopoda* roots, resulting in no change in vegetative growth.

MATERIALS & METHODS—*Study area*. This study was conducted on McKenzie Flats in the Sevilleta National Wildlife Refuge (SNWR), located 80 km south of Albuquerque, New Mexico, USA (34°20′37.73″ N, 106°43′38.45″ W). Within the SNWR, the study site lies in Chihuahuan Desert grassland dominated by *B. eriopoda* (Poaceae), which is found on homogeneous soils (Buxbaum & Vanderbilt 2007). *Bouteloua eriopoda* is a warm-weather, stoloniferous perennial grass and typically thrives at elevations of 3,500 to 5,500 ft (U.S. Department of Agriculture Natural Resources Conservation Service [NRCS] 2005). New vegetation grows from basal axillary buds and from axillary buds at nodes on stolons (NRCS 2005). The climate at the SNWR is semi-arid with an average annual temperature of 13.2°C, mean low of 1.6°C in January and high of 25.4°C in July (Thomey et al. 2011; Báez et al. 2013). Summer monsoon rainstorms occur July–September with an average annual precipitation of 250 mm (Thomey et al. 2011; Báez et al. 2013).

Experimental design. To test the effects of an extreme precipitation treatment on *B. eriopoda*, I conducted a rainfall manipulation experiment for four weeks during June and July 2018 in a semi-arid grassland. Fourteen 1 m^2 plots were randomly selected from a 40 x 15 m area located 30 m east of a dirt road. Seven plots received a precipitation treatment of 10 mm of water two times per week for four weeks, while seven plots received no treatment. This represented above-average monsoon events, in contrast to the average of < 5 mm of water per event (Petrie et al. 2014). The water treatment was applied with a watering can to a 1.2 m^2 area so that the water treatment for each 1 m^2 plot extended 0.2 m beyond each plot, assuring that the plant rooting zone within plots received the same amount of water throughout. Treatment water was obtained through reverse-osmosis and contained 0.67 ppm NO^{3-}-N and 1.01 ppm NH^{4+}-N (Thomey et al. 2011). I chose to apply 10 mm of water to each plot since few precipitation inputs (11.6%) are between 10 and 20 mm in the SNWR (Thomey et al. 2011; Petrie et al. 2014). Although 10 mm rainstorms are not the most extreme precipitation events to this ecosystem, the frequency of this treatment (biweekly) is not common and thus, is considered extreme (Petrie et al. 2014). This treatment simulated increases in both rainfall frequency and size. Response variables included vegetative growth, soil moisture, and soil stability.

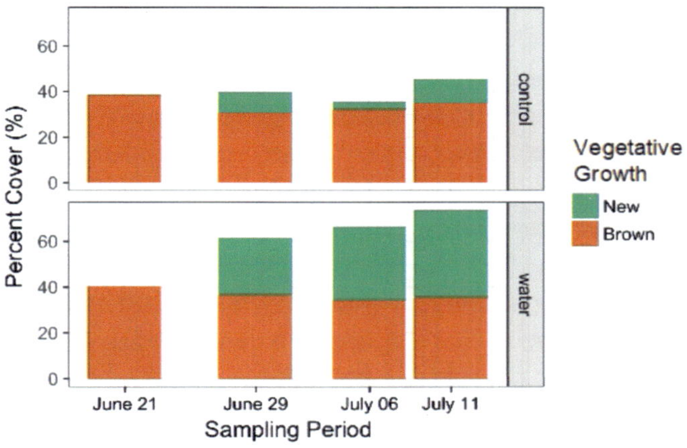

FIGURE 1. Vegetative growth was sampled in 14 plots, 7 control and 7 watered, within one site in *Bouteloua eriopoda*-dominated grassland in the Northern Chihuahuan Desert, New Mexico, USA. Vegetative growth was sampled four times throughout the experiment. Percent cover of new foliage and brown foliage of *B. eriopoda* was recorded during each sampling date when present. Brown foliage remained constant across all sampling dates for control and water (t-stat = -1.56, df = 48, $P = 0.12$). New foliage differed between watered and control plots (t-stat = -13.0, df = 32, $P < 0.001$). Total cover (new and brown) differed between the treatment and control plots across all sampling dates (t-stat = -5.13, df = 65, $P < 0.001$).

Foliar cover and vegetative growth were measured once per week over four weeks to track responses of *B. eriopoda* to increased rainfall. Foliar cover was determined by estimating the percent of 1 dm^2 quadrant squares occupied by each plant. Vegetative growth was measured by visually estimating the percent of new and brown foliage from the total percent cover, which was derived from the foliar cover measurements. Before treatment began, *B. eriopoda* accounted for 30–50% of total cover within all plots. Other subdominant species included *Gutierrezia sarothrae*, *Plantago patagonica*, and *Chamaesyce lata*. Vegetative growth was only described for *B. eriopoda* within the 1 m^2 plots because this dominant grass displayed the most growth and cover change compared to all other species during this experiment.

Soil stability was not correlated with vegetative growth of *B. eriopoda*, so this study compared soil stability between short-term and long-term experiments. Soil stability was quantified at three different sites located on McKenzie Flats, SNWR: (i) a short-term monsoon experiment (this study); (ii) a long-term monsoon experiment (Monsoon Rainfall Manipulation Experiment, MRME); and (iii) a long-term drought experiment (Extreme Drought in Grassland Experiment, EDGE). Monsoon Rainfall Manipulation Experiment was located 52 m E from this experiment and EDGE was located 703 m SW from this experiment. Soil stability was tested and ranked using the method described by Herrick et al. (2015) in the Monitoring Manual for Grassland, Shrubland and Savanna Ecosystems. Soil stability was ranked on a scale from 1 to 6 after dipping the soil sample in water. A ranking of 1 (low stability) indicates that < 10% of the soil sample remained after 30 sec after dipping the sample in water five times. A ranking of 6 (high stability) indicates that 75–100% of the soil sample remained after five water dips.

Soil moisture was measured as volumetric water content (%VW) using a handheld soil moisture probe, Moisture Meter type HH2 (Delta-T Devices Ltd., Cambridge, UK). Soil moisture was measured immediately before and after each watering treatment and the consecutive 1–2 days after each watering treatment. Soil moisture measurements were recorded at the edge and center of each plot at a depth of 16 cm, within the recognized rooting zone of *B. eriopoda* (Thomey et al. 2011). The edge and center soil moisture measurements were averaged for each plot before analysis.

Data analysis. Data analysis was completed in Microsoft Excel (version 16.15, Redmond, WA, USA). Two-sample t-tests were used to compare the differences in vegetative cover between control and watered plots. These t-tests analyzed brown and new foliage separately to find differences in those variables between control and watered plots (FIG 1). Analysis of variance (ANOVA) with replication was used to analyze the total foliar cover between the first and last sampling dates, June 21 and July 11, for control and watered plots (TABLE 1). The α was set to 0.05 for all analyses. All figures were created in R, version 3.5.0 (R Core Team 2018).

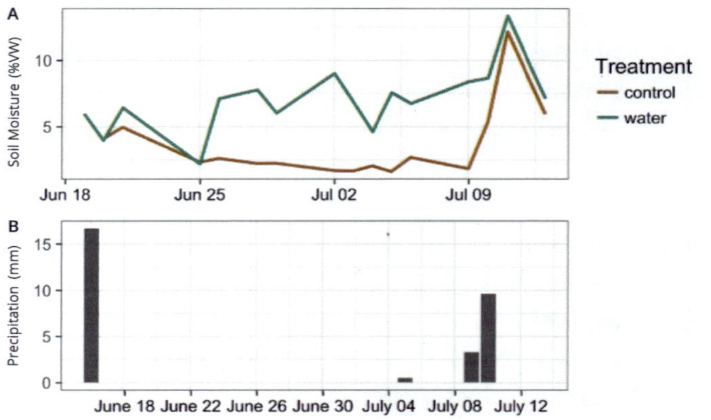

FIGURE 2. Daily record of soil moisture content and natural precipitation throughout the four-week simulated monsoon treatment period during June–July 2018. (A) Average soil moisture (% VW) was recorded at 160 mm depth for each water and control plot and then averaged across each treatment. (B) Two natural rain events occurred on 16 Jun 2018 and 10 Jul 2018.

RESULTS—*Change in vegetative growth over time*. This study was influenced by two ambient rain pulses. The first ambient rain pulse occurred on 16 Jun 2018, before the first sampling date, and the second ambient rain pulse occurred on 9–10 Jul 2018, before the last sampling date (FIG 2). On the first sampling date, 21 Jun 2018, soil moisture (%VW) was similar between control and watered plots, which averaged between 5–6%VW. Throughout the experiment, watered plots maintained the same average soil moisture while control plots had low soil moisture of 2%VW. On the last sampling date, 11 Jul, both control and watered plots increased soil moisture to 8 and 10%VW, respectively, due to the second ambient rain pulse. Throughout this experiment, brown

TABLE 1. Analysis of total foliar cover between control and watered plots of the first (21 Jun 2018) and last (11 Jul 2018) sampling days. Calculations were determined with a two-way ANOVA with replication.

Source of Variation	SS	df	MS	*F*	*P*-value	*F* crit
Sample	2735.37	1	2735.38	28.9615	1.5687×10^{-5}	4.2597
Columns	1586.27	1	1586.27	16.7950	0.0004	4.2597
Interaction	1192.77	1	1192.77	12.6288	0.0016	4.2597
Within	2266.77	24	94.4487			
Total	*7781.19*	*27*				

foliage remained constant in both watered and control treatments (t-stat = -1.56, df = 48, $P = 0.12$). On the second vegetative growth sampling day, 29 Jun, both control and watered plots showed new foliage on all *B. eriopoda* plants. Percent cover of new foliage increased slightly in watered plots for the remainder of the experiment (FIG 1 & 3). New foliage increased in control plots on the last vegetative growth sampling day, which occurred one day after the second natural rain-pulse to the experimental site (FIG 1, 3, & 4). Total foliar cover in both treatment and control plots increased over the course of the experiment (TABLE 1; $F = 28.96$, df = 1, $P < 0.001$), although the total cover of watered and control plots differed at the end of the experiment (TABLE 1; $F = 12.63$, df = 1, $P = 0.0016$). Change in vegetative growth and total foliage cover were not correlated with soil stability.

Soil stability at three sites. Soil stability was measured at three different sites to determine the differences in soil stability between a short-term monsoon treatment, a long-term monsoon treatment, and a long-term drought treatment in *B. eriopoda*-dominated grassland. This experiment, a short-term monsoon treatment, found no difference in soil stability between watered and control treatments (FIG 5; t-stat =0.15, df = 11, $P = 0.89$). In MRME, a long-term monsoon experiment, there were no significant differences among all treatments (FIG 5; $F = 1.17$, df = 2, $P = 0.35$). In EDGE, a long-term drought experiment, soil stability was significantly higher in control plots compared to drought treatment (FIG 5; t-stat = 5.66, df = 11, $P = 0.0001$).

DISCUSSION—*Change in vegetative growth over time*. The results of this study support the hypothesis that an extreme precipitation treatment will have a positive effect on total foliar cover and vegetative growth change of *B. eriopoda*. Total foliar cover and new foliage increased in all plots. There were two natural rain pulses that influenced the study system, one on 16 Jun and the other on 9–10 Jul (FIG 2 & 3). The first natural rain pulse caused a noticeable change in vegetative growth in both treatments on the second sampling day, 29 Jun 2018 (FIG 1). There is some discrepancy in FIG 2 where the heavy spike in ambient rainfall on 16 Jun did not increase soil moisture to a similar degree as the ambient rainfall on 9 Jul did. Since precipitation data were collected from a rain gauge located at MRME, 53 m from this study site, there may have been unequal rainfall to this experimental site since monsoons are known to unevenly distribute rain across the landscape due to localized cells (Petrie et al. 2014). The second natural rain pulse caused a more noticeable effect in control plots than watered plots (FIG 3 & 4). Even though watered plots maintained higher total cover than control (t-stat = -5.13, df = 65, $P < 0.001$), control plots responded slightly faster to the second natural rain pulse than did watered plots (FIG 4). This can be related

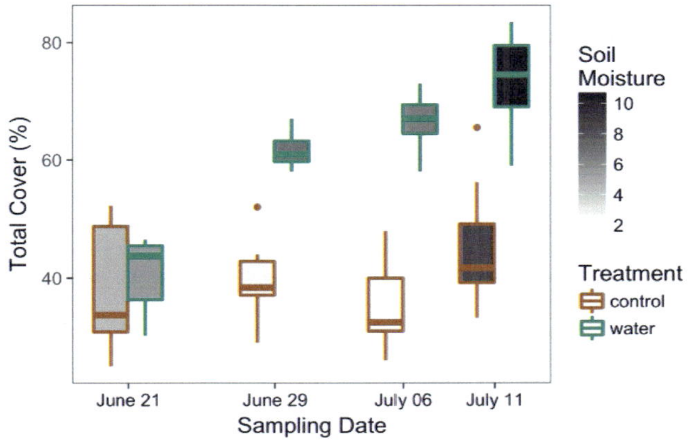

FIGURE 3. Comparison of total foliage cover of *B. eriopoda* between watered and control plots across four sampling periods. Brown box plots represent control plots that received ambient rainfall. Teal box plots represent watered plots that received 10 mm water biweekly plus ambient rainfall. Shading within each box plot represents the average soil moisture (%VW) of each treatment, white for low soil moisture and black for high soil moisture.

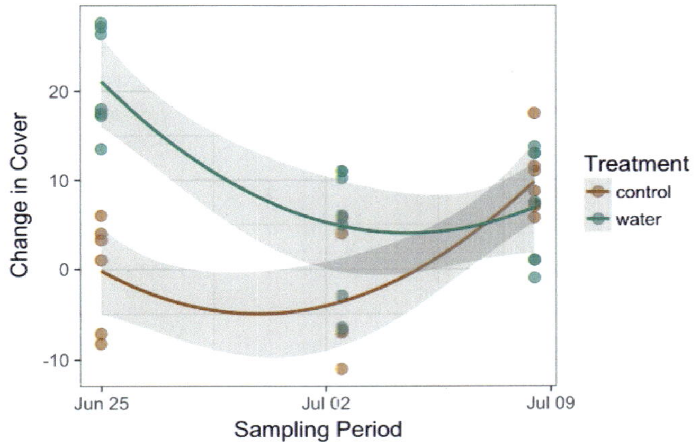

FIGURE 4. Change in cover of *B. eriopoda* compared among four sampling periods. Change in cover was calculated for each consecutive sampling period and is shown for each mid-date between each sampling period. The blue line represents plots that received weekly watering treatments and the red line represents control plots that received no treatment. Points represent cover measured for each individual plot. Upper and lower limits of best fit lines are ± SE.

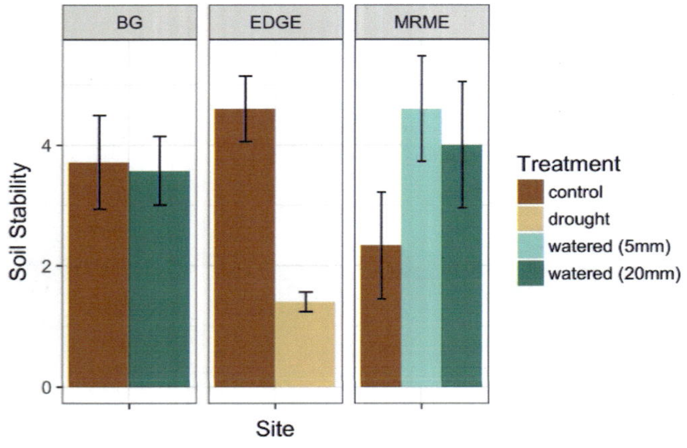

FIGURE 5. Comparisons of soil stability at three different sites in *B. eriopoda*-dominated grassland in the Northern Chihuahuan Desert, New Mexico, USA. The sites are labeled as follows: short-term monsoon treatment (BG), long-term drought treatment (EDGE), and long-term monsoon treatment (MRME). Legend labels are associated with separate treatments within each site. In BG, soil stability did not differ between ambient rainfall and 10 mm biweekly waterings (t-stat = 0.15, df = 11, $P = 0.89$). In EDGE, soil stability significantly differed between ambient rainfall and drought treatment (t-stat = 5.66, df = 11, $P = 0.0001$). In MRME, soil stability did not differ between ambient rainfall and the two watering treatments ($F = 1.17$, df = 2, $P = 0.35$). Soil stability is ranked on a scale from 1 to 6; 1 is low stability and 6 is high soil stability. Error bars are ± SE.

to previous findings at MRME where *B. eriopoda* responded more favorably to rain pulses that are separated by dry periods (20 mm water/mo) rather than continuous watering cycles (5 mm water/wk; Thomey et al. 2011).

These results show the importance of coupling short-term experiments with long-term experiments. Even though long-term studies like MRME (Thomey et al. 2011) and EDGE (unpublished data) found that *B. eriopoda* responds more favorably to rain pulses separated by dry periods across the growing season, short-term experiments are able to examine the more immediate responses of *B. eriopoda*. Oftentimes, long-term experiments only focus on species composition or aboveground primary productivity. This broad focus of long-term experiments may show that plants respond to climate change, but they do not focus on the underlying mechanisms of plant responses, such as the rapid foliar response of *B. eriopoda*, as this short-term experiment demonstrated. Referring to Ogle & Reynolds' (2004) elaboration on pulse-dynamics, the watered plots were probably operating near their maximum growth rate threshold whereas control plots were operating near their minimum growth rate threshold (FIG 4). This is inferred by the slightly higher increase and steeper trajectory of percent cover in control plots compared to watered plots (FIG 4). Longer monitoring of this experiment, coupled with natural rainfall events, would lead to a more deterministic conclusion.

Soil stability at three sites. The results do not support the hypothesis that increasing water to the system would increase soil stability. Soil stability was not correlated to soil moisture, nor the change in *B. eriopoda* vegetative growth. Since soil stability was not correlated with vegetative growth in this experiment, soil stability was analyzed at a long-term monsoon experiment (MRME) and a long-term drought experiment (EDGE) to determine differences between short-term treatments and long-term treatments. This experiment and MRME showed no difference in soil stability between their respective treatments, whereas EDGE did have significant differences in soil stability between treatments. The soil stability results from MRME are most likely due to small sampling size and a range of soil stability results within each treatment. However, there is an important trend within MRME. I found that the soil was slightly more stable within the 5 mm water/wk plots compared to the 20 mm water/mo plots. Since biocrusts are sensitive to changes in pulse sizes (Fernandes at al. 2018), biocrusts may favor consistent soil moisture contrary to *B. eriopoda*, which had more biomass at the 20 mm water/mo plots at MRME (Thomey et al. 2011), although further experimentation is needed to draw a more solid conclusion. When comparing soil stability rankings from MRME to this study, the watered treatments at MRME had higher soil stability than either treatment in this experiment.

CONCLUSIONS—Because climate change is predicted to lead to an increase in the magnitude and frequency of wet and dry years (Knapp et al. 2015), understanding how semi-arid plants respond to precipitation extremes is pertinent to how plant dynamics may be altered in the future. This study showed that precipitation extremes increase the total cover and new foliage of *B. eriopoda*, though the magnitude of that change depends on the amount of precipitation *B. eriopoda* receives (FIG 1, 3, & 4). Soil stability differed between short-term and long-term precipitation manipulation experiments. Long-term experiments had more stable soils, suggesting that long-term treatments may be necessary for sustaining growth of biocrusts (Fernandes et al. 2018). This study is a stepping stone for long-term experiments to determine the effects of climate change in a semi-arid ecosystem. Although there was no relationship between total foliar change and soil stability, long-term experiments can incorporate a range of soil stability testing to determine that relationship, such as analyzing soil stability on a gradient from open interspaces to under plant canopies. Future investigation of the minute changes to semi-arid grasslands due to extreme precipitation will help expand our understanding of how semi-arid ecosystems will change within a varying arid climate.

ACKNOWLEDGMENTS—This project was funded by a grant from the National Science Foundation for the Sevilleta Research Experience for Undergraduates program presented by the University of New Mexico. The Fish and Wildlife Service approved this research at the Sevilleta National Wildlife Refuge. University of New Mexico provided vehicles and supplies for this project, managed by Stephanie Baker. Thanks to Renée F. Brown, Alesia Hallmark, Stephanie Baker, and Scott Collins who provided guidance for this study. Renée F. Brown assisted with project design and comments that greatly improved this manuscript. Alesia Hallmark assisted with data analysis and graph design using R. Any errors are my own and should not tarnish the reputation of any of these esteemed persons.

REFERENCES

Adair, E. C., Burke, I. C. 2010. Plant phenology and life span influence soil pool dynamics: *Bromus tectorum* invasion of perennial C_3–C_4 grass communities. *Plant Soil* 335: 255–269.

Báez, S., Collins, S. L., Pockman, W. T., Johnson, J. E., Small, E. E. 2013. Effects of experimental rainfall manipulations on Chihuahuan Desert grassland and shrubland plant communities. *Oecologia* 172: 1117–1127.

Buxbaum, C. A. Z., Vanderbilt, K. 2007. Soil heterogeneity and the distribution of desert and steppe plant species across a desert-grassland ecotone. *Journal of Arid Environments* 69: 617–632.

Chaudhary, V. B., Bowker, M. A., O'Dell, T. E., Grace, J. B., Redman, A. E., Rillig, M. C., Johnson, N. C. 2009. Untangling the biological contributions to soil stability in semiarid shrublands. *Ecological Applications* 19: 110–122.

Collins, S. L., Belnap, J., Grimm, N. B., Rudgers, J. A., Dahm, C. N., D'Odorico, P., Litvak, M., Natvig, D. O., Peters, D. C., Pockman, W. T., Sinsabaugh, R. L., Wolf, B. O. 2014. A multiscale, hierarchical model of pulse dynamics in arid-land ecosystems. *Annual Review of Ecology, Evolution, and Systematics* 45: 397–419.

Collins, S. L., Sinsabaugh, R. L., Crenshaw, C., Green, L. , Porras-Alfaro, A., Stursova, M., Zeglin, L. H. 2008. Pulse dynamics and microbial processes in aridland ecosystems. *Journal of Ecology* 96: 413–420.

Eldridge, D. J., Greene, R. S. B. 1994. Microbiotic soil crusts: A review of their roles in soil and ecological processes in the rangelands of Australia. *Australian Journal of Soil Research* 32: 389–415.

Fernandes, V. M. C., Machado de Lima, N. M., Roush, D., Rudgers, J., Collins, S. L., Garcia-Pichel, F. 2018. Exposure to predicted precipitation patterns decreases population size and alters community structure of cyanobacteria in biological soil crusts from the Chihuahuan Desert. *Environmental Microbiology* 20: 259–269.

Ferrenberg, S., Reed, S. C., Belnap, J. 2015. Climate change and physical disturbance cause similar community shifts in biological soil crusts. *Proceedings of the National Academy of Sciences of the United States of America* 112: 12116–12121.

Herrick, J. E., Van Zee, J. W., McCord, S. E., Courtright, E. M., Karl, J. W., Burkett, L. M. 2015. *Monitoring Manual for Grassland, Shrubland and Savanna Ecosystems*, Second Edition, Volume I: Core Methods. Las Cruces, NM: USDA-ARS Jornada Experimental Range. pp. 45–47

Knapp, A. K., Avolio, M. L., Beier, C., Carroll, C. J. W., Collins, S. L., Dukes, J. S., Fraser, L. H., Griffin-Nolan, R. J., Hoover, D. L., Jentsch, A., Loik, M. E., Phillips, R. P., Post, A. K., Sala, O. E., Slette, I. J., Yahdjian, L., Smith, M. D. 2017. Pushing precipitation to the extremes in distributed experiments: recommendations for simulating wet and dry years. *Global Change Biology* 23: 1774–1782.

Knapp, A. K., Hoover, D. L., Wilcox, K. R., Avolio, M. L., Koerner, S. E., La Pierre, K. J., Loik, M. E., Sala, Y. O. E., Smith, M. D. 2015. Characterizing differences in precipitation regimes of extreme wet and dry years: Implications for climate change experiments. *Global Change Biology* 21: 2624–2633.

Noy-Meir, I. 1973. Desert ecosystems: Environment and producers. *Annual Review of Ecology and Systematics* 4: 25–52.

Ogle, K., Reynolds, J. F. 2004. Plant responses to precipitation in desert ecosystems: Integrating functional types, pulses, thresholds, and delays. *Oecologia* 141: 282–294.

Petrie, M. D., Collins, S. L., Gutzler, D. S., Moore, D. M. 2014. Regional trends and local variability in monsoon precipitation in the northern Chihuahuan Desert, USA. *Journal of Arid Environments* 103: 63–70.

R Core Team. 2018. R: A language and environment for statistical computing. R Foundation for Statistical Computing, Vienna, Austria. URL https://www.R-project.org/.

Reynolds, J. F., Kemp, P. R., Ogle, K., Fernández, R. J. 2004. Modifying the "pulse-reserve" paradigm for deserts of North America: Precipitation pulses, soil water, and plant responses. *Oecologia* 141: 194–210.

Thomey, M. L., Collins, S. L., Friggens, M. T., Brown, R. F., Pockman, W. T. 2014. Effects of monsoon precipitation variability on the physiological response of two dominant C_4 grasses across a semiarid ecotone. *Oecologia* 176: 751–762.

Thomey, M. L., Collins, S. L., Vargas, R., Johnson, J. E., Brown, R. F., Natvig, D. O., Friggens, M. T. 2011. Effect of precipitation variability on net primary production and soil respiration in a Chihuahuan Desert grassland. *Global Change Biology* 17: 1505–1515.

U.S. Department of Agriculture Natural Resources Conservation Service. 2005. Black Grama [Fact Sheet]. Retrieved from https://plants.usda.gov/factsheet/pdf/fs_boer4.pdf.

Wang, J., Wu, C., Zhang, C., Ju, W., Wang, X., Chen, Z., Fang, B. 2018. Improved modeling of gross primary productivity (GPP) by better representation of plant phenological indicators from remote sensing using a process model. *Ecological Indicators* 88: 332–340.

Diversity and Abundance of Soil Microbes Differ Along a Forest-Pasture Transect

Hannah Suli[1*], Ashley Schumann[1], Cleo Bickley[1], Jasmine Rodriguez[1]

KEYWORDS—*microbial diversity, microbes, bacteria, ecology, land management, soil diversity, soil health, ecosystem functioning*

INTRODUCTION—Soil microbes are instrumental in orchestrating the overall function and diversity of soil communities through a multitude of ecological processes, including carbon cycling, nitrogen cycling, nutrient acquisition, and soil structuring (Van Der Heijden et al., 2007). Preliminary studies reveal that changes in the abundance of soil organisms causes marked increases or decreases in ecosystem functional diversity, elucidating the importance of microbes in driving ecosystem productivity (Rich et al., 2003; Wagg et al., 2014). Despite all of the known roles that microbes play in soil ecology, the degree to which plants influence microbe diversity remains unclear. We are only in the initial stages of understanding the extent of symbiotic relationships between microbes and plants, illuminating the need for further examination of soil ecology. A better understanding of these interactions could potentially improve our assessment and management of agricultural or disturbed settings. We sought to quantify the spatial variation of soil microbe communities along a gradient spanning from a densely forested area, intersecting an equine trail, and ending in an adjacent fallow pasture. We hypothesized that due to heightened plant diversity in the forest, there would be higher microbial abundance and diversity in the forest than in the pasture and that microbial abundance and diversity will decrease along the transect, from forest to pasture.

MATERIALS & METHODS—We conducted our study in the Dows Prairie area of McKinleyville, California, USA. The pasture, previously grazed by cattle and left out to fallow for at least 10 years, borders a horse trail and adjacent forest. The forest is mostly undisturbed except for foot traffic from local wildlife and horseback riders on marked trails. We collected soil samples with a sterilized stainless-steel spoon every 6.1 m (20 ft) along a 121.9 m (400 ft) transect (FIG 1). We placed samples in resealable plastic bags and stored them in a freezer overnight. To examine bacterial diversity, we plated two cultures per sample, each with 1 x 10^{-5} g of serially diluted soil. We accomplished this by placing 1 g of soil into a test tube containing 9.9 mL saline solution, vortexing the test tube, and transferring 100 µL into another test tube containing 9.9 mL saline solution. Lastly, we plated 100 µL of the final dilution on agar enriched with lysogeny broth. To examine fungal diversity, we placed a small sample of undiluted soil on water agar.

We left our samples to culture in a box at room temperature. Twenty-four hours following plating, we recorded the number of distinct colonies per plate and

*Corresponding Author: hms66@humboldt.edu
[1]Undergraduate Student
Department of Biological Sciences
Humboldt State University, Arcata, CA 95521

Suli, H., Schumann, A., Bickley, C., Rodriguez, J. 2019. Diversity and abundance of soil microbes differ along a forest-pasture transect. *ideaFest Journal: Interdisciplinary Journal of Creative Works & Research from Humboldt State University* 3: 36–39.

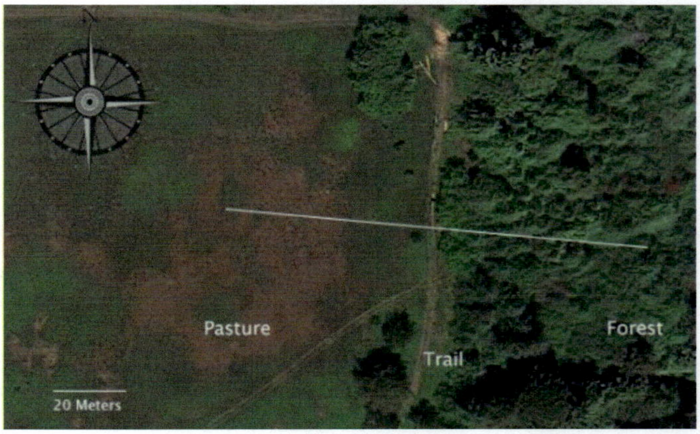

FIGURE 1. Map of Dows Prairie study area, with transect in white, intersected by a horse trail.

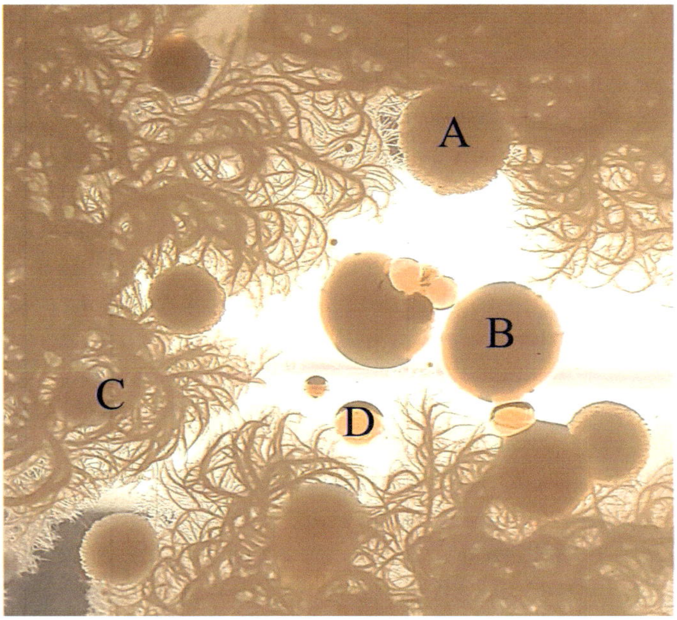

FIGURE 2. Example bacteria colony morphotypes, cultured from a pasture soil sample. (A) Grainy, (B) white, (C) filamentous, and (D) transparent.

sorted each based on phenotype (see FIG 2 for examples). We repeated our observations and colony counts every 24 hr over the course of three days. Filamentous bacteria versus fungal phenotypes were distinguished with the help of a bacteriologist and senior researchers at Humboldt State University.

We focused our statistical analysis on data from our final day of data collection (day three), when distinct bacterial colonies and their morphotypes for all plates were most easily observable. After recording our raw data, we analyzed the effect of distance from forest canopy to open pasture on observed bacterial colony abundance (n = 20) using linear regression. We determined the effect of location either within or outside of the dense forest canopy on colony abundance with a Wilcoxon rank sum test. We analyzed the effects of both distance and location within the forest or pasture on Shannon Diversity (H') and morphotype richness (R) using a Welch 2-sample t-test and a Wilcoxon rank sum test, respectively. We calculated diversity indices for final observations on day three using an online calculator (Goepel, 2012) and conducted all statistical analyses using RStudio (2015).

RESULTS—Distance from canopy did not affect bacteria colony abundance ($R^2 = 0.032$, $P = 0.14$; FIG 3). However, we did observe that abundance was greater overall in the pasture than in the forest (W = 104.5, $P = 0.01$; FIG 4). We observed a mean of 90.2 ± 7.96 bacteria colonies from pasture samples and 62.35 ± 9.46 bacteria colonies from forest samples. Shannon indices (H') were dependent on distance from canopy ($t = 7.14$, df = 19, $P < 0.0001$) and were significantly higher in the pasture than in the forest (W = 8, $P < 0.0001$; FIG 5). Distance from canopy affected morphotype richness per sample ($t = 7.02$, df = 19, $P < 0.0001$), but richness did not differ significantly between the pasture and the forest (W = 32.5, $P = 0.18$; FIG 6). We observed a mean of 4.7 ± 0.15 morphotypes in pasture samples and 3.7 ± 0.09 morphotypes in forest samples.

DISCUSSION & CONCLUSIONS—Our data did not support our initial hypothesis that microbial abundance or

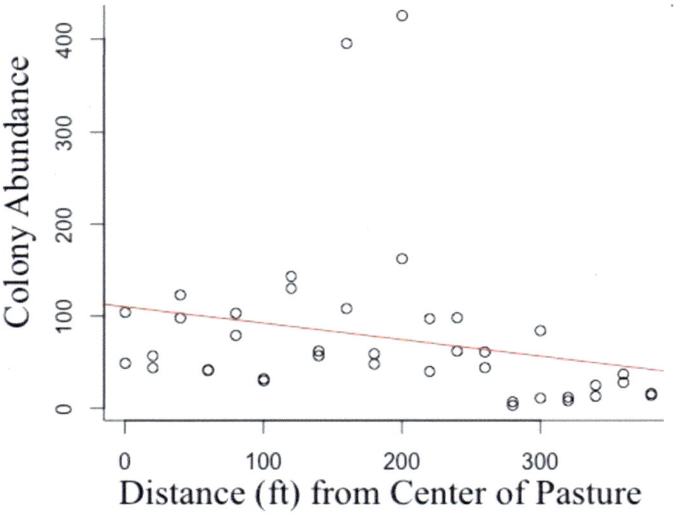

FIGURE 3. Total bacteria colonies per plate as a function of location along the transect, with regression line ($R^2 = 0.032$).

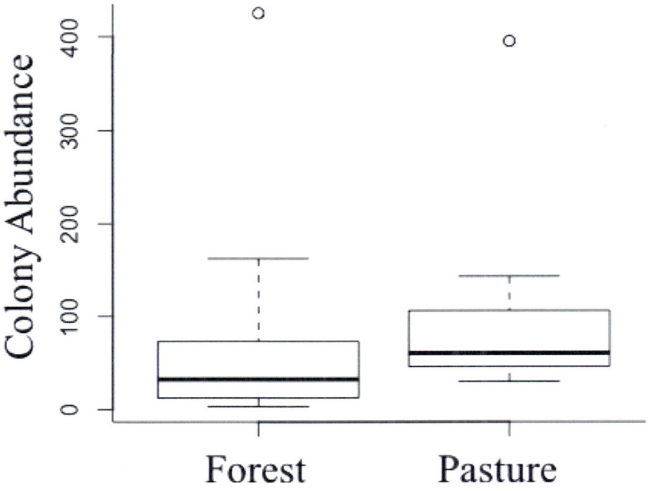

FIGURE 4. Distribution of colony abundance in forest versus pasture. There was a significant difference in colony abundance between forest and pasture samples ($P = 0.01$).

diversity would decrease with distance from forest. The data did, however, suggest an inverse relationship for diversity. Diversity was significantly higher in pasture samples than in forest samples, and both richness and Shannon indices increased along the transect (FIG 5, FIG 6). Colony abundance was not a function of distance from canopy (FIG 3), but abundance was greater overall in pasture soil (FIG 4). The increase in richness across the transect from the forest into the pasture indicates that there is a larger distribution of bacterial morphospecies in the pastoral environment than in the forested environment.

The fact that soil bacterial abundance was not affected by distance from canopy, but differed between the two conditions, might suggest a greater microbial survival

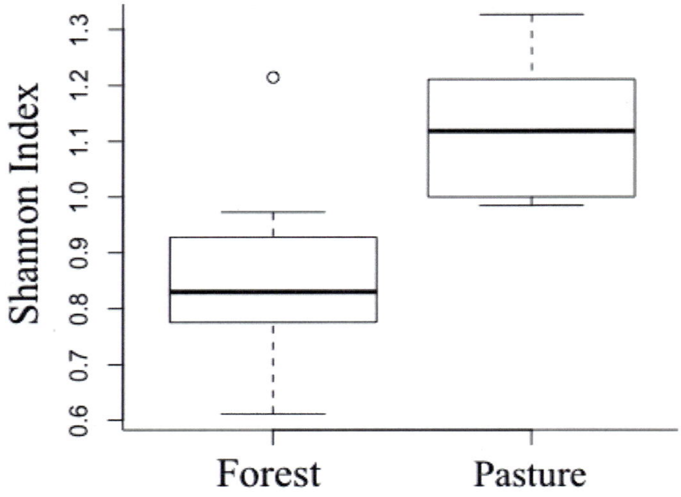

FIGURE 5. Comparison of average Shannon indices in forest and pasture.($P < 0.001$).

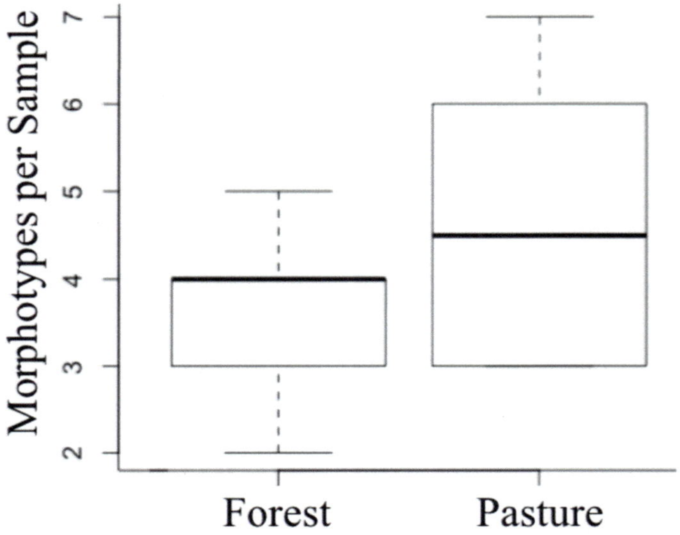

FIGURE 6. Comparison of average morphotype richness between forest and pasture ($P = 0.18$).

relationship with root contact or organic matter than with temperature or shade variations caused by vicinity to canopy. Bacteria often exhibit symbiotic relationships with plants, and while we expected greater microbial abundance and diversity in the forest, weedy root presence in pasture samples could explain why the opposite occurred (Yang, 2009). Grass and shrub roots occur at a much shallower depth than the roots of mature trees, meaning plant contact with surface soils was more consistent in our pasture samples.

It is unclear how, and to what extent, equine traffic contributed to greater bacterial abundance and diversity in the pasture. Bacteria from intentional application of manure in agricultural settings can survive weeks or months, even in unsuitable environments, and manure is known to change the physical properties of soil, including pH and ionic concentrations (Fenlon et al., 2000; Unc et al., 2003). Foot and large animal traffic in the forest is limited by high plant and tree density, even at outer forest boundaries, but traffic along the trail at the transect's midpoint is frequent, and manure deposits likely contributed to bacterial communities in close contact with the trail or pasture. Furthermore, a mild downward slope from the trail to the center of the pasture could redirect rain flow, and thus the flow of equine-related bacteria. This could explain two high-abundance outliers near the trail (FIG 3) in both pasture and forest samples, and could explain why morphotype richness was dependent on distance, but did not necessarily differ significantly between pasture and forest (FIG 6). Ideally, we would either repeat this experiment in a setting not intersected by trails, or we would analyze trafficked soil samples separately in their own category. Regardless, our data suggest that microbial analysis offers a promising measure of animal impact on soil health.

Limited lab availability meant that our forest samples were cultured one day later than our pasture samples, even though they were collected on the same day. Prolonged exposure to cold temperatures might have reduced the viability of bacteria from the forest, skewing diversity results in favor of the pasture. A second limitation of our experiment was that we were not able to gather enough data to quantify fungal diversity. No recognizable fungal colonies grew on our water agar plates, which were dominated by bacteria. In the future we would culture fungal samples in a media more conducive to fungal growth alone. Only one bacteria colony formed on an open plate, out of two closed control plates and two control plates that remained open during the duration of sample plating, indicating that contamination was not a likely factor in our results.

Lack of bacterial taxonomic knowledge hindered identification of our culture colonies. As such, we focused on observed phenotypes rather than species identification. Certain morphotypes, such as filamentous or deep red bacteria, were easily distinguishable, but others manifested only subtle differences in transparency or texture (FIG 2). Given more time and resources, we would perform DNA extractions and sequencing on each colony morphotype in order to gain a more complete picture of represented species and their ecological significance. A carbon source utilization test would be beneficial in distinguishing the functional diversity of bacteria in forested versus pastoral environments. Given the time constraints of our study, we did not have the opportunity to conduct a survey of the plant species present in the forested and open areas. Future studies would include diversity indices for all present plant species to account for the interaction between plant diversity and bacterial abundance.

ACKNOWLEDGMENTS—This research was supported by the College of Natural Resources Core Research Facility at Humboldt State University, with guidance from David Baston and Tim McClure, and materials provided by Terilyn Stoflet.

REFERENCES

Fenlon, D., Lewis, D., Leach, K., Howard, R., Svoboda, I., & Ogden, I. 2002. Fate of *Escherichia coli* and *Escherichia coli* O157 in soils and drainage water following cattle slurry application at 3 sites in southern Scotland. *Soil Use & Management* 18: 223–231.

Goepel, K. D. 2012, October. BPMSG Diversity online calculator. Retrieved from https://bpmsg.com/academic/div-calc.php.

RStudio Team. 2015. RStudio: Integrated development for R. RStudio, Inc., Boston, MA. URL http://www.rstudio.com/.

Rich, J. J., Heichen, R. S., Bottomley, P. J., Cromack, K., Myrold, D. D. 2003. Community composition and functioning of denitrifying bacteria from adjacent meadow and forest soils. *Applied & Environmental Microbiology* 69: 5974–5982.

Wagg, C., Bender, S., Widmer, F., van der Heijen, A. 2014. Soil biodiversity and soil community composition determine ecosystem multifunctionality. *Proceedings of the National Academy of Sciences of the United States of America* 111: 5266–5270.

Unc, A., Goss, M. J. 2004. Transport of bacteria from manure and protection of water resources. *Applied Soil Ecology* 25: 1–18.

van der Heijden, M., Bardgett, R., van Staalen, N. 2007. The unseen majority: Soil microbes as drivers of plant diversity and productivity in terrestrial ecosystems. *Ecology Letters* 11: 296–310.

Yang, J., Kloepper, J. W., Ryu, C. 2009. Rhizosphere bacteria help plants tolerate abiotic stress. *Trends in Plant Science* 14: 1–4.

The Effects of Citric Acid on pH and Nutrient Uptake in *Triticum aestivum*

Cristina L. Tusei[1*]

KEYWORDS—*citric acid, chelate, wheatgrass, soil contamination, nutrient uptake*

INTRODUCTION—Citric acid is a metabolite in plants and plays an important role in photosynthesis and cellular respiration. It contains three carboxylic acid (COOH) functional groups (FIG 1) and can deprotonate to form a negatively charged citrate molecule.

The loss of external hydrogen atoms enhances citric acid's ability to act as a chelate and form complexes around positively charged ions such as metals, effectively removing these cations from solution. Citric acid has a different binding constant for various cations (Goli et al. 2012). The molecule's difference in affinity for positively charged molecules affects which metals it will complex and the rate at which this process will happen. This is an important concept to consider when studying the removal of heavy metals from the environment, a prominent environmental concern. Anthropogenic activities have increased the presence of elements that are harmful to plants and animals in concentrations above threshold levels (Hussain et al. 2019), and methods for their removal are currently being researched. A novel technique called phytoremediation has become a method of interest to decrease the concentrations of heavy metals in the environment. Phytoremediation is the ability of plants to sequester select compounds by secreting a chelating agent; the plant is then able to metabolically convert the element in question into something less harmful (Gong et al. 2018). Prior experiments suggest that citric acid aids elemental uptake in plants by forming a complex around the cation in question, whether it be beneficial or potentially harmful (Senden et al. 1994; FIG 2). This effectively increases both essential and non-essential nutrient availability to plants (Zhang et al. 2018) but also carries the potential for toxicity to occur within plant tissues if the threshold dose level is surpassed. Plants require macronutrients such as calcium (Ca), magnesium (Mg), and potassium (K) for biochemical processes, as well as micronutrients like copper and iron. Their ability to uptake metals also means they will draw harmful elements such as lead and cadmium through their roots, which can be detrimental after prolonged exposure. Metallic species present in the environment in close proximity to vegetation and wildlife often depend on soil composition and surrounding establishments. However, the general trend suggests that areas situated by increasing urbanization and industrial processes have experienced an increase in the concentrations of such metals. Various studies on removing detrimental elements from the environment have been conducted and phytoremediation shows promising results (Gong et al. 2018).

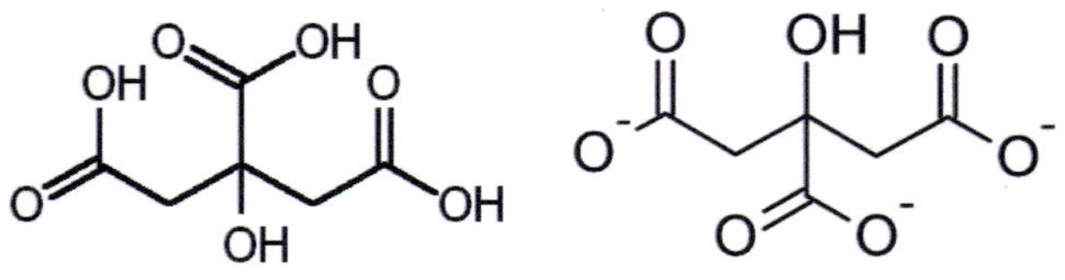

FIGURE 1. A comparison of a citric acid molecule (A) and citrate molecule (B).

*Corresponding Author: clt19@humboldt.edu
[1]Undergraduate Student
Department of Biological Sciences
Humboldt State University, Arcata, CA 95521

Tusei, C. L. 2019. The effects of citric acid on pH and nutrient uptake in *Triticum aestivum*. *ideaFest Journal: Interdisciplinary Journal of Creative Works & Research from Humboldt State University* 3: 40–44.

Assorted chelating agents, such as ethylenediaminetetraacetic acid (EDTA) and citric acid, have been utilized in phytoremediation research. A key component in determining functional ligands is examining their effects on the plants themselves. For example, EDTA is effective in complexing metals but is toxic to plants and biodegrades slowly (Hussain et al. 2019). Citric acid

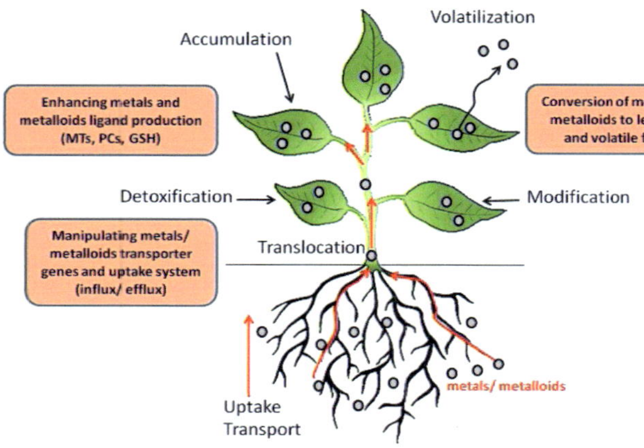

FIGURE 2. The transport of metals through a plant and their conversion to a less toxic form (from Mosa et al. 2016).

degrades in the environment naturally and is excreted by plants themselves, so it is suggested that using this molecule is more beneficial than utilizing other chemical species (Hussain et al. 2019). The use of plants in removing harmful elements from the environment is a fairly new and promising technology. However, because it can take years to restore a landscape by phytoremediation, it can only be used in sites that have low to medium contaminant concentrations. The effectiveness of phytoremediation also depends on the plant species and the heavy metals present in the environment.

Triticum aestivum, or common wheat from the family Poaceae, is a widespread plant that has established itself nationwide and proliferates in all seasons. It is highly tolerant in various environments and grows rapidly, making it the ideal plant to study. Due to citric acid's ability to chelate metals and be absorbed by plants, it is hypothesized that exposure to it will increase the ability of wheatgrass to absorb macro- and micro-nutrients, as well as heavy metals, from soil. This experiment will specifically examine the elements calcium (Ca), magnesium (Mg), potassium (K), and nitrogen (N); certain limitations of this

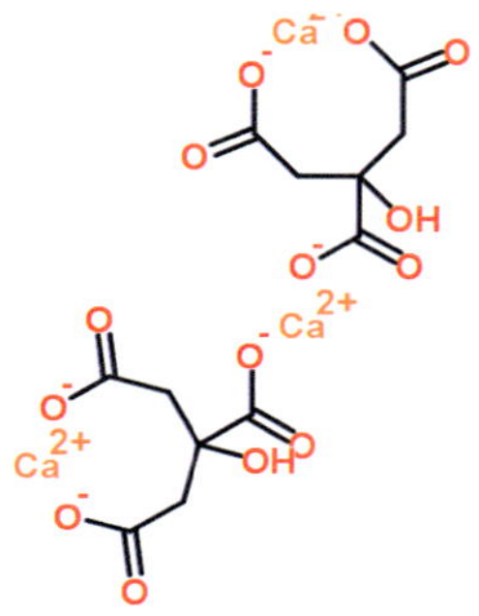

FIGURE 3. Citrate as chelate for calcium

inquiry made the application of heavy metals to solution not possible. The aforementioned nutrients are vital to plant growth and biochemical processes and results obtained can be used to predict how heavy metals such as lead and mercury would be absorbed by a plant similar to wheatgrass. Given that citric acid has a pH of 2.2, it is also hypothesized that prolonged exposure of wheatgrass to the organic compound will increase the acidity of the soil.

METHODS—*Triticum aestivum*, also known as common wheatgrass, was selected as the test subject due to its versatility in growing conditions and rapid growth rate. Seeds were purchased from Pierson's Garden Care Center in Eureka and germinated in Humboldt State University's experimental greenhouse. Six seeds were sowed per pot, with two replicate pots for each of four treatments and two replicate pots for the control. Plants were kept in a naturally lit area and were grown in a standard potting mix containing compost, perlite, alfalfa, and crushed basalt, among other nutrient-containing materials. The control group received 200 mL of deionized water (diH_2O) per pot. Treatment groups also received 200 mL of diH_2O per pot in addition to the following concentrations of citric acid: 5 mM, 10 mM, 15 mM, and 20 mM. Each replicate contained six plants and treatments were administered three times a week for three months.

Chemical assays. The pH was measured every two weeks initially, then increased to once a week after a month. Results were obtained by capturing water drainage from the soil and analyzing the solution with a pH meter. Macronutrients Ca, Mg, and K were analyzed using an Atomic Absorption Spectrometer (AAS). Solutions for AAS were prepared by drying and ashing plant material and filtering with 20% hydrochloric acid (HCl). After being treated with HCl, samples were diluted to 50 mL with diH_2O and centrifuged. Further serial dilutions were made with 2% HCl in order to obtain the proper range of readings on the AAS; otherwise the concentrations of each element would have been inaccurate. Serial dilutions were factored into calculating the concentration of each element.

Nitrogen content was measured using Kjeldahl digestion and titration, in which organic N was converted into ammonia following a reaction with sulfuric acid during a digestion period. The solution was reacted with 40% sodium hydroxide and distilled into 50 mL of a boric acid solution containing bromocresol green, methyl red, 100% ethanol, p-nitrophenol, and 4% boric acid. This solution

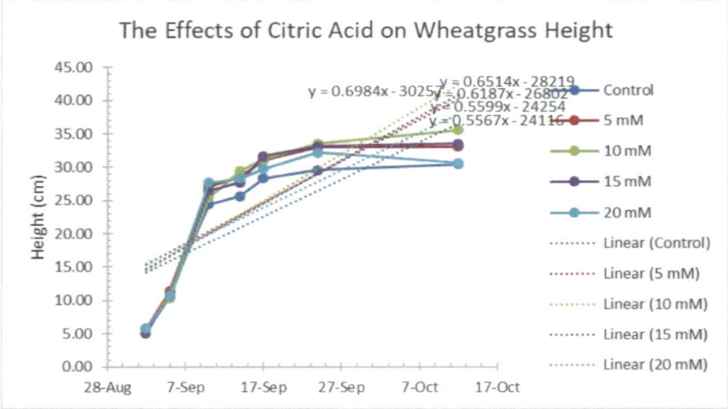

FIGURE 4. The effects of citric acid on wheatgrass height.

was then titrated with 0.1 M HCl until the endpoint was reached, indicated by an orange-red color.

RESULTS—An initial increase in height occurred for all treatments, including the control. The most efficient way to determine if one treatment grew more than another was through computing the slope of each line; the steeper the slope, the greater the growth rate of that particular treatment. Examining the slopes, it is clear that neither the control nor any of the treatments decreased in height and that the 10 mM wheatgrass experienced the largest growth rate, with the steepest slope of 0.6984. The control, with a slope of 0.5567, experienced the lowest growth rate.

The pH of the treatment groups steadily decreased (FIG 6). The control group varied so it is possible that other factors besides the addition of citric acid affected the pH of the soil. Observed decreases in pH suggest that citric acid may have caused a drop in pH; this was also visually apparent during the experiment, as water outflow was noticeably paler with increasing citric acid concentration (FIG 5). Statistical analyses were limited to R^2 values that were calculated based on linear modeling of pH over time. All treatments, including the control, had R^2 values greater than 0.50, indicating that these data fit the model better than by chance. In particular, the 10 mM treatment had an R^2 value of 0.807, indicating that this treatment best fit the model, in which there was a linear decrease of pH during the course of the experiment. The 5 mM treatment also had a high R^2 value ($R^2 = 0.769$), indicating that this treatment also experienced a decrease in pH over time. While I cannot comment on the significance of these effects, these effects were strong for 5 and 10 mM treatments in particular.

FIG 7 displays the effects that varying molarities of

FIGURE 5. A visual representation of the effects of citric acid on the outflow of water from wheatgrass plants. Molarity increases from left to right.

citric acid had on the concentrations of Ca, Mg, and K. Calcium decreased slightly at 10 mM and again at 20 mM. Potassium had a sharp dip at 15 mM but increased again at 20 mM; magnesium stayed constant. As indicated by the very low R^2 values for these linear models, the strength of the effect of citric acid concentration on the concentration of Ca, Mg, and K was extremely weak.

FIG 8 portrays the concentration of N analyzed through a Kjeldahl digestion. The percentage of N present in approximately one gram of sample was consistently higher in citric acid treatments relative to the control. Given an R^2 value of 0.4847, the data do not closely fit the linear model of increasing percentage of N with increasing citric acid concentration.

DISCUSSION—Plants often secrete organic acids from their roots in order to complex free metallic ions and draw the element up to their shoots (Wang et al. 2017). This includes both beneficial and harmful chemical species that can impact the plant's metabolic processes depending on the concentration of the element absorbed. Factors that determine metal uptake include the species of plant and the type of organic acid present. It is suggested that plants are able to convert toxic metals into nontoxic forms (Mosa et al. 2016). This occurs to an extent, until the metals become toxic to the plant itself and adversely affect it.

Favored chelating agents include EDTA, nitriloacetate

(NTA), diethylenetriaminepentaacetate (DTPA), and, more recently, citric acid (Hasegawa 2012; FIG 9). Biodegradable compounds such as citric acid are utilized more often due to the fact that they will break down in the environment instead of persisting like the aforementioned chemicals.

This study hypothesized that after being exposed to increasing increments of citric acid concentration, wheatgrass plants would contain higher amounts of macronutrient elements and the soil in which they reside would have a lower pH. The results obtained suggest that pH will decrease as molarity increases; however, this cannot be supported statistically and can only be inferred from FIG 5 and R^2 values. There is no strong evidence to support that nutrient metal uptake is influenced by varying molarities of citric acid.

Although results of this experiment were inconclusive, inquiries of the same nature have been conducted elsewhere with promising results. A study concerning the nutrient profile of wheatgrass at Marathwada Agricultural University in Parbhani, India, suggests that macronutrient concentrations increase linearly over time (Kulkarni et al. 2006). The study analyzed a total of fifteen elements by use of instrumental neuron activation analysis. Wheatgrass was grown for a total of twenty days in three different treatments: tap water, tap water with nutrients, and soil and tap water. All treatments were watered with 200 mL and kept in an unfiltered air and light setting. The main differences between this experiment and the referenced study lie in the varying treatments the plants were placed in and the amount of water they received. It is a possibility that a solution already treated with nutrients would increase the likelihood of there being a greater concentration of the nutrients within the plant already. While it was anticipated that macronutrient concentration would increase over time, I did not observe this in this study. This may be due to other contributing factors that were not incorporated into the initial reasoning, such as physiological strain or interactions with other chemical species.

The methods used in this experiment can be improved upon. The trial was run for two months, and this may not have been sufficient time to observe a difference in the effects of citric acid on elemental absorption. A more productive treatment window would have been six months to a year. The plants were grown in a greenhouse, but no special precautions, such as temperature and light control, were incorporated. It is possible results would have been significant had these factors been accounted for. Instrumental error could have also occurred but replicates of the analyses would need to be carried out in order to determine this. In the future, this experiment should be performed for a longer amount of time within an enclosed environment. Multiple replicates of instrumental

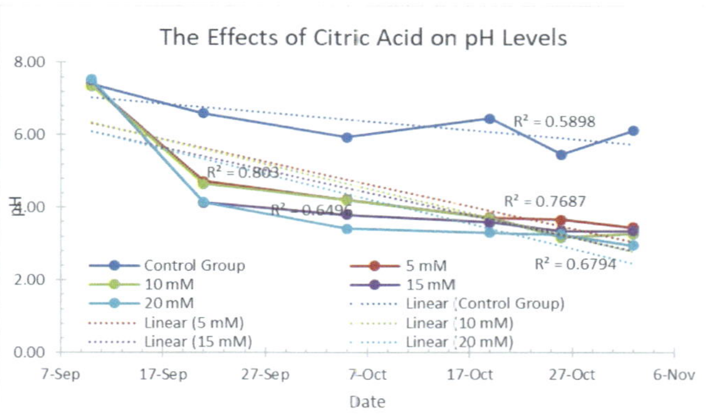

FIGURE 6. The effects of citric acid on pH.

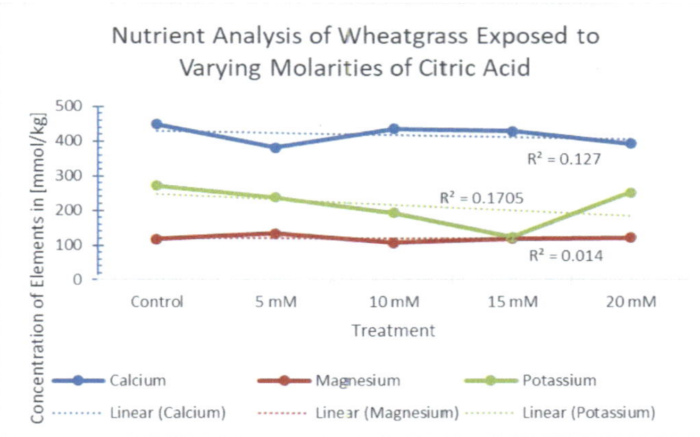

FIGURE 7. The variation in the amounts of nutrients in mmol/kg for each treatment of citric acid.

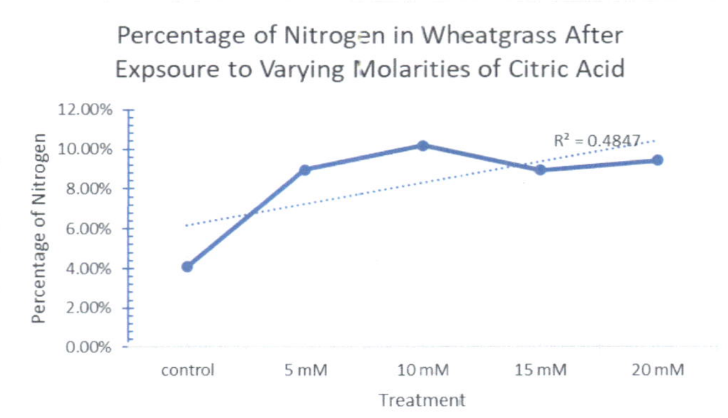

FIGURE 8. Percentage of nitrogen present in varying molarities of citric acid.

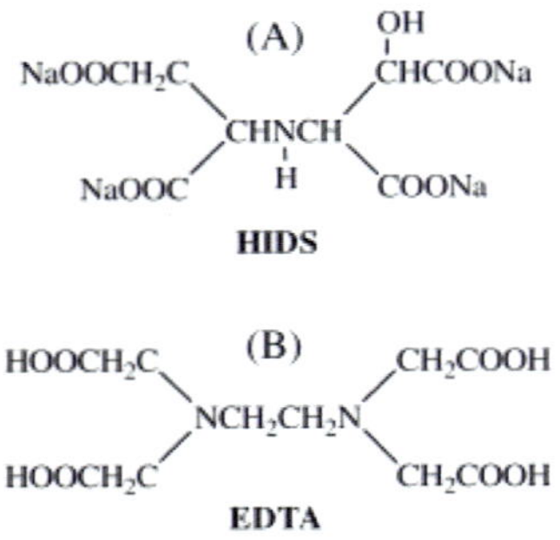

FIGURE 9. The chemical structures of two chelating compounds, HIDS and EDTA (from Hasegawa 2012).

analyses should also be performed with a greater number of treatment pots.

This experiment suggested that addition of citric acid to solution will decrease soil pH. Results are not conclusive as to whether it affects elemental absorption and thus it is difficult to ascertain its ability to improve phytoremediation ability of plants. Had results been more conclusive, the utilization of citric acid as a chelate and its ability to remove metals from solution would have been an indication for further experiments to determine its effectiveness in phytoremediation. It is recommended that future studies be conducted in order to better determine the effect of citric acid on metal and nutrient absorption in plants. Possible experimental improvements include prolonging the treatment period, growing plants in more controlled conditions, and spiking each treatment with a different nutrient element, such as Mg or Ca, and analyzing the plant for that particular metal. Solidified outcomes hold implications for future treatment of plants in nutrient deficient or contaminated soils.

ACKNOWLEDGMENTS—Many thanks to Dr. Casey Lu, who supervised this study and provided the procedures for the AAS and Kjeldahl chemical assays. A big thank you to Dr. Dave Baston, for continued support and use of the Core Research Center. I would also like to express my appreciation for Susan Wright, who allowed the use of select equipment and aided in waste removal.

REFERENCES

Angumeenal, A. R., Venkappayya, D. 2013. An Overview of Citric Acid Production. *LWT – Food Science and Technology* 50: 367–370.

Camaselle, C., Gouveia, S. 2018. Phytoremediation of mixed contaminated soil enhanced with electric current. *Journal of Hazardous Materials* 361: 95–102.

Coe, F. Calcium Binding by Citrate, The University of Chicago, [online image]. https://kidneystones.uchicago.edu/calcium-binding-by-citrate.

Goli, M. B., Pande, M., Bellaloui, N. 2012. Effects of chelating agents on protein, oil, fatty acids, and minerals in soybean seed. *Agricultural Sciences* 3: 517–523.

Gong, X., Huang, D., Liu, Y., Zeng, G., Wang, R., Wei, J., Huang, C., Xu, P., Wan, J., Chen, Z. 2018. Pyrolysis and reutilization of plant residues after phytoremediation of heavy metals contaminated sediments: For heavy metals stabilization and dye adsorption. *Biosource Technology* 253: 64–71.

Hasegawa, H., Rahman, M. M., Kadohashi, K., Takasugi, Y., Tate, Y., Maki, T., Rahman, M. A. 2012. Significance of the concentration of chelating ligands on Fe^{3+} solubility, bioavailability, and uptake in rice plant. *Plant Physiology & Biochemistry* 58: 205–211.

Hussain, A., Kamran, M. A., Javed, M. T., Hayat, K., Farooq, M. A., Ali, N., Ali, M., Mangwhar, H., Jan, F., Chaudhary, H. J. 2019. Individual and combinatorial application of *Kocuria rhizophiliai* and citric acid on phytoextraction of multi-metal contaminated soils by *Glycine max* L. *Environmental & Experimental Botany* 159: 23–33.

Kulkarni, S. D., Archaya, R., Nair, A. G. C., Rajurkar, N. S., Reddy, A. V. R. 2006. Determination of elemental concentration profiles in tender wheatgrass (*Triticum aestivum* L.) using instrumental neutron activation analysis. *Food Chemistry* 95: 699–707.

Mosa, K. A., Saadoun, I., Kumar, K., Helmy, M., Dhankher, O. P. 2016. Potential biotechnological strategies for the cleanup of heavy metals and metalloids. *Frontiers in Plant Science* 7: 303.

Mourato, M. P., Moreira, I. N., Leitao, I., Pinto, F. R., Sales, J. R., Martins, L. L. 2015. Effect of heavy metals in plants of the genus *Brassica*. *International Journal of Molecular Sciences* 16: 17975–17998.

Senden, M. H. M. N., van der Meer, A. J. G. M., Verburg, T. G., Wolterbeek, H. T. 1995. Citric acid in tomato plant roots and its effect on cadmium uptake and distribution. *Plant & Soil* 171: 333–339.

Wang, S., Dong, Q., Wang, Z. 2017. Differential effects of citric acid on cadmium uptake and accumulation between tall fescue and Kentucky bluegrass. *Ecotoxicology & Environmental Safety* 145: 200–206.

Zhang, X., Zhong, B., Shafi, M., Guo, J., Liu, C., Guo, H., Peng, D., Wang, Y., Liu, D. 2018. Effect of EDTA and citric acid on absorption of heavy metals and growth of moso bamboo. *Environmental Science & Pollution Research* 25: 18846–188852.

Determining Fire Severity of the 2017 Santa Rosa, CA Fire

John Cortenbach[1], Richard Williams[1], Buddhika Madurapperuma[2*]

ABSTRACT—This study examines the 2017 Santa Rosa wildfire using remote sensing techniques to estimate the acreage of burned areas. Landsat 8 imagery of the pre- and post-fire areas was used to extrapolate the burn severity using two methods: (i) difference Normalized Burn Ratio (dNBR) and (ii) change detection analysis. We compared our analyses with data provided by the California Department of Forestry and Fire Protection (Cal Fire). The results of burn severity using both methods were on average 24% under-approximated in comparison to Cal Fire values. When comparing our acreage burn analysis to Cal Fire data, our results were on average 76 ± 8% accurate in identifying burn severity. Of the two methods, the change detection using an iso-clustered unsupervised classification scheme was more accurate. Landsat-based burn severity mapping provides cost–effective tools for forest managers to identify fire risk areas for wildfire planning and ecosystem management.

INTRODUCTION—Catastrophic wildfire is among the most common form of forest disturbance in the Pacific Northwest of the United States. In California and adjacent states there is fear that the recent spate of high-intensity fires that wipe out canopy trees, or stand-replacing fires, present a challenge for conservation of conifer forests. In the beginning of the 2017 fall season in northern California, large fires erupted and spread throughout Santa Rosa, Napa, and Sonoma Counties (Miller, 2017). The scars left by these wildfires, though terrible, provided an opportunity to investigate the impact they had on the affected areas mentioned above. Wildfires in California are frequent due to climatic conditions that are influenced by dry fuel accumulation, lightning, lengthening of the summer drought, and earlier spring snowmelt and runoff (Stewart et al., 2005; Westerling et al., 2006; Williams et al., 2010). It is suspected that high temperatures and fast winds, characteristic of the fall season, fueled the rapid spread of the wildfire across counties in northern California.

In 2017, a multitude of fire outbreaks occurred throughout the state of California. One of the most extreme was the Santa Rosa fire across Sonoma and Napa Counties, which put many homes at risk and rendered the air quality hazardous throughout much of that summer. These fires caused severe damage to structures and destroyed many wine-cultivating areas.

*Corresponding Author: bdm280@humboldt.edu
[1]Alumnus
Department of Environmental Science and Management
Humboldt State University, Arcata, CA 95521
[2]Lecturer/Research Associate
Departments of Forestry and Wildland Resources and Environmental Science and Management
Humboldt State University, Arcata, CA 95521

Cortenbach, J., Williams, R., Madurapperuma, B. 2019. Determining fire severity of the 2017 Santa Rosa, CA fire. *ideaFest Journal: Interdisciplinary Journal of Creative Works & Research from Humboldt State University* 3: 45–52.

Tubbs and Nuns fire. The 2017 wildfires in northern California burned through and destroyed whole neighborhoods in and adjacent to the city of Santa Rosa, Sonoma County (Madsen et al., 2018). Two blazes, namely the Tubbs and Nuns fires, ravaged huge swaths of Sonoma County. The Tubbs fire began on 8 Oct 2017 and burned portions of Napa, Sonoma, and Lake Counties during October that year. It was estimated that over 36,807 acres had burned by 31 Oct 2017. The escalation of the fire was heavily attributed to the peak wind gusts ranging between 20 and 40 miles per hour that pushed the fire from the northeast. According to Watkins et al. (2017), the fire spread approximately 12 miles in the first three hours of its outbreak. The Nuns fire, on the other hand, was located east and north of the city of Sonoma and spread to both Sonoma and Napa Counties. This fire also began on

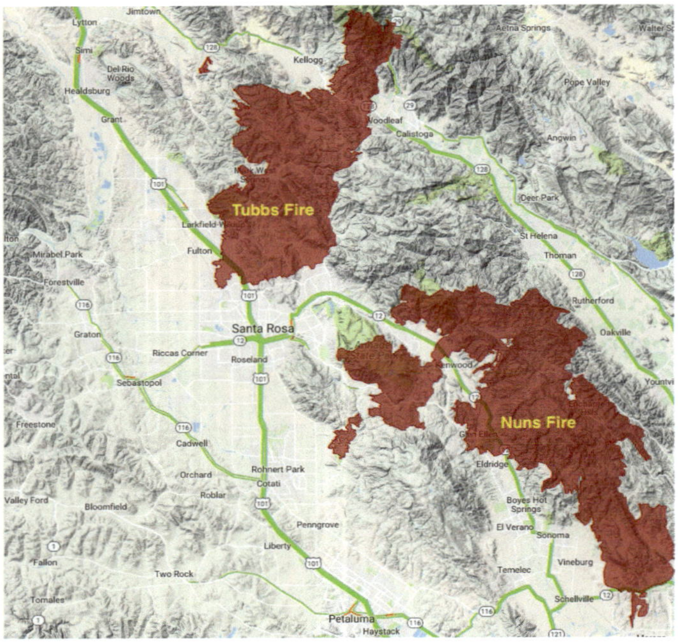

FIGURE 1. Google map of the Tubbs and Nuns fires after their destructive barrage over Sonoma County (Nor Cal Fires, n.d.).

8 Oct 2017 and spread across an estimated 54,000 acres. On its spread path, it merged with other fires, such as the Adobe, Pressley, and Norrbom fires, absorbing them and creating one massive fire (Watkins et al., 2017). Due to limited fire outbreak datasets, this study was constrained and dependent on two pre- and post-fire Landsat datasets of the Tubbs and Nuns fires that were collected in Nov 2017 (FIG 1). It is important to note that these two images did not span the same region of interest since the two fires did not overlap (FIG 1).

The scope of this study is to determine the severity of the Tubbs and Nuns fires by estimating the acreage of burned areas within Sonoma County using the following specific objectives: (i) determine which fire had the highest severity by comparing the difference Normalized Burn Ratio (dNBR) between the Tubbs and Nuns fires; (ii) estimate the burn acreage within the two fire regions with change detection using a pixel-over-pixel comparison of an iso-clustered unsupervised classification; and (iii) estimate the actual burned acreage values and a degree of error associated with the estimation of burned acreage.

METHODS—*Delineate burned area*. Due to the recent occurrence of the northern California fires, specifically in Santa Rosa, the preliminary analysis was limited to 2017 Landsat 8 imagery that was obtained on 9 Sept and 5 Nov 2017 from the United States Geological Survey (USGS, 2017). These two images reflected the pre- and post-fire states, respectively. FIG 2 displays the Landsat images used for this analysis with true color band combinations used to enhance the contrast (band 6, 5, 4 combinations for red, green, and blue, respectively). These three band combinations were of shortwave one and two composite (SWIR1, SWIR1, SWIR2) and a SWIR1, near infrared (IR), and red bands. Such burned areas are dependent on an accurate region of interest (ROI), which is then created in near IR and red bands. These bands are assigned to the blue, green, and red bands. These band combinations helped to discriminate burned and unburned areas, hence creating an ROI.

After the ROI was identified, as shown in FIG 2, the overall workflow used to achieve the objectives consisted of two methods: (i) difference Normalized Burn Ratio (dNBR) and (ii) change detection using an iso-cluster unsupervised classification. The workflow for these analyses is presented in FIG S1.

difference Normalized Burn Ratio (dNBR). In image preprocessing, digital numbers were converted to Top of Atmosphere (ToA) reflectance using ArcMap© (Rozario et al., 2018). Dark object subtraction was performed to remove the effect of atmospheric scattering on the reflectance of dark objects, such as water pixels, using ENVI® (Rozario et al., 2018). The reflectance bands of the pre- and post-fire images were used to derive Normalized Burn Ratio (NBR).

The NBR formula uses near IR (NIR) and the shortwave infrared (SWIR-2) wavelengths as shown in Equation (1).

$$\mathrm{NBR} = (\mathrm{NIR} - \mathrm{SWIR}) \,/\, (\mathrm{NIR} + \mathrm{SWIR}) \qquad (1)$$

Then, difference Normalized Burn Ratio (dNBR) was computed by subtracting the post-fire NBR from the pre-fire NBR as shown in Equation (2).

$$\mathrm{dNBR} = \mathrm{NBR}_{pre} - \mathrm{NBR}_{post} \qquad (2)$$

The dNBR analysis was conducted partially in ENVI®, to manipulate bands, create ROIs, and clean up the data. The images were then imported into ArcMap© to run the NBR indexing for the pre- and post-fires images, followed by the dNBR.

Change detection. ENVI's Image Change workflow was

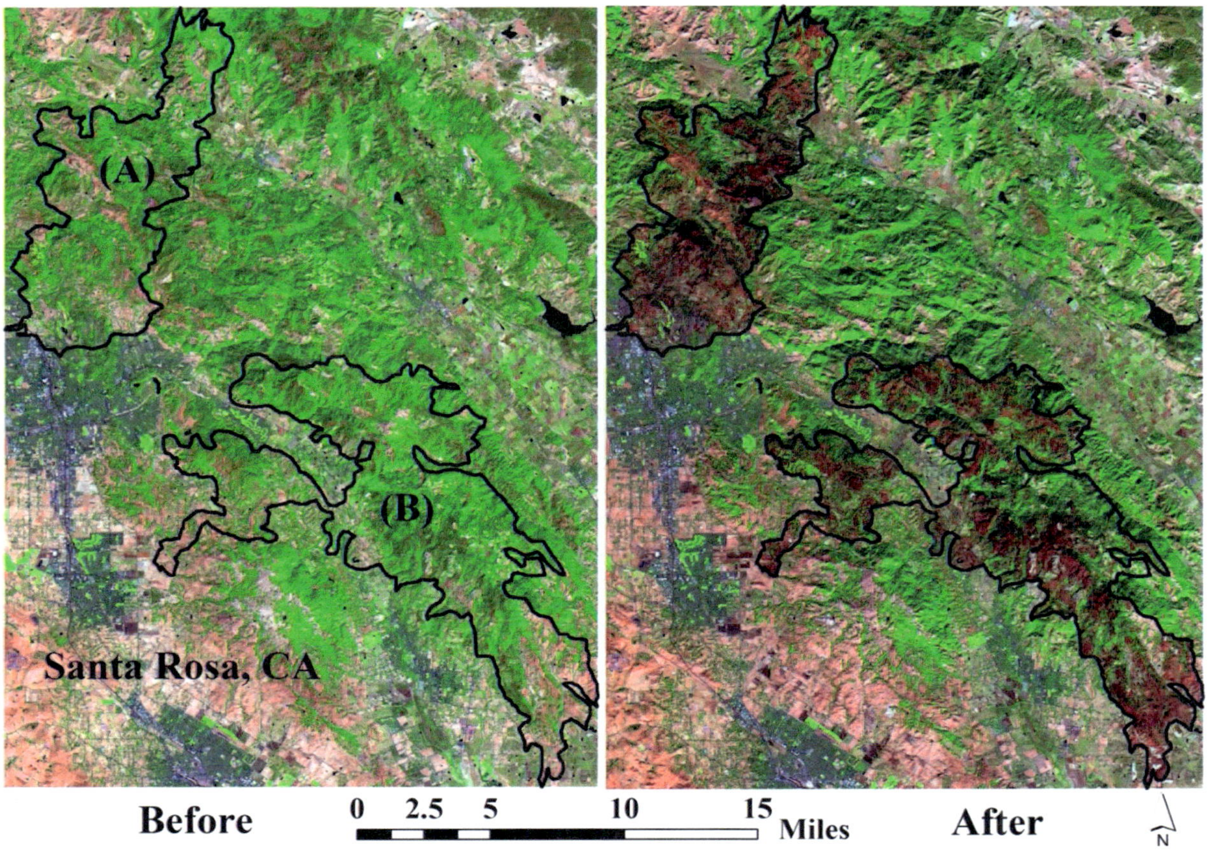

FIGURE 2. Santa Rosa, California 2017 burn severity study area pre-fire (left) and post-fire (right). Tubbs (A) and Nuns (B) fires are to the north and west of the city, respectively. The image appearance is set to a true color 6, 5, 4 Landsat band combination. The black outline ROI indicates the area used for the analysis.

used to determine burn areas (Exelis, 2015). The Image Change workflow compares spectral differences of two images of the same geographic region and different time steps, and also identifies differences between them. The difference can be computed on a thermal band corresponding to the temperature difference between pre- and post-fire. Burn extent and severity classes were created by taking the difference between pre- and post-NBR layers.

We estimated burned acreage using dNBR and change detection methods. Since the area estimation from the above methods was based on remote sensing methods, we utilized Cal Fire burn acreage data to compare the results. Cal Fire estimated burned acreage by compiling various dataset shared from partners such as USDA Forest Service, Bureau of Land Management, National Park Service, US Fish and Wildlife Service, and numerous local agencies (https://frap.fire.ca.gov/data/statewide/FGDC_metadata/fire18_1_metadata.xml). Various data sources of Cal Fire resulted in multi-methods for burn acreage estimation.

RESULTS—The dNBRs for both the Tubbs and Nuns fires were analyzed and presented with respect to the city of Santa Rosa, as shown in TABLE 1 and FIG 3. Qualitatively it appears that the severity between the Tubbs and Nuns fires for both dNBR and change detection estimation methods is more or less similar. However, through close visual inspection of the zoomed-in images, it appears that there are more high-severity zones present in the Nuns ROI than the Tubbs area. From this dNBR analysis of the Landsat images, the severity files were reclassified into specific areas associated with low, medium, and high burn severities.

Both methods were very close in their approximations,

TABLE 1. Comparison of the approximated burned acreage between the Tubbs and Nuns fires using the dNBR and a change detection method.

Parameter	dNBR (acres)		Change Detection (acres)	
	Nuns	Tubbs	Nuns	Tubbs
Estimated	37,503	23,998	38,355	24,579
Actual	46,104	35,270	46,104	35,270
Accuracy	81%	68%	83%	70%

but were on average 24% under-approximated when compared to values supplied by California Department of Forestry and Fire Protection (Cal Fire, 2017). The errors in classification are detectable in FIG 3 where there are unburned areas in close proximity to high-severity burn areas. The shapefiles used to determine the areas, which were converted from raster to vector files, are shown in FIG 4. This image is overlaid on the burn areas so that overlap and non-overlap of the shapefiles can visually show where burned areas were not categorized. The imagery noise correction analysis, like reflectance, cloud cover, and air moisture, greatly improve the accuracy of the analysis.

A comparison of the two fires, with respect to burn severity, was conducted using ground truthing data to validate the remote sensing results (FIG 5). A field visit to Santa Rosa fire areas was made in mid-March 2018 to observe the impact of the fire as well as to validate our burn severity classification maps. A total of 60 ground-truthing points were tested for burn severity and 45 points were matched with dNBR analysis, resulting in 75% accuracy for classifying and references images.

FIG 6 summarizes the burn acreage of both Nuns and Tubbs fires in 2017 in Santa Rosa. These results showed that the Nuns fire had more total acreage burned than the Tubbs fire. However, the percentage burn for low, medium, and high burn severity classes for both fires was more or less similar. For example, of the total area of the Nuns and Tubbs fires, 3% and 4% were high severity, 22% and 25% were moderately burn, and 76% and 71% were low burn, respectively. Using Cal Fire records, it was identified that our analysis was moderately accurate (76 ± 8%) (Jenson, 2005).

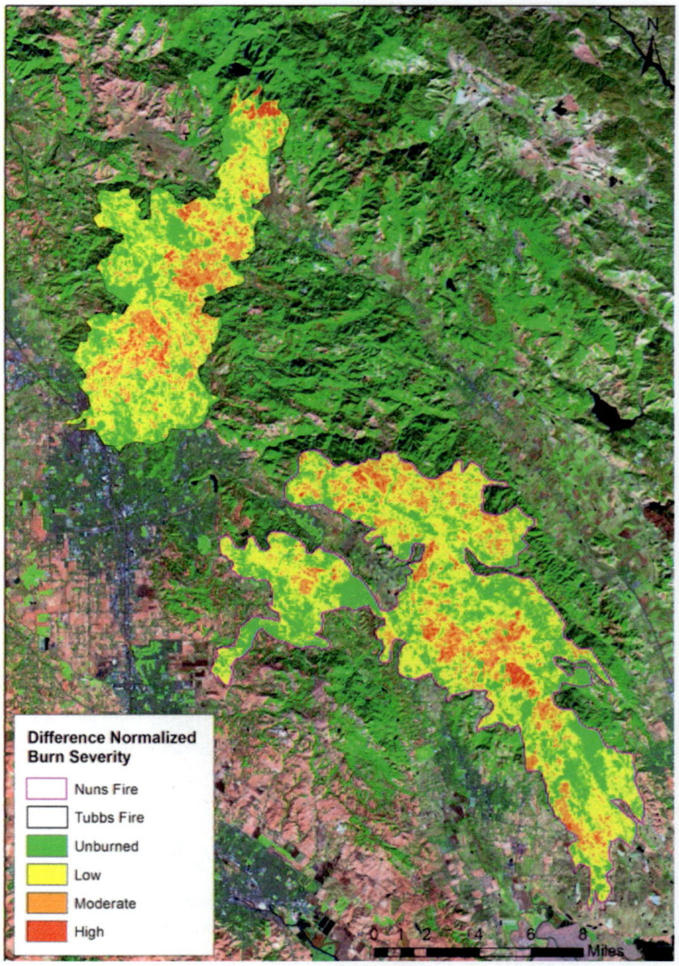

FIGURE 3. dNBR analysis for the Tubbs fire (top-left) and Nuns fire (bottom-right) using four main classes to categorize burn severity in the greater Sonoma and Napa County areas.

DISCUSSION—Wildfire in California has increased dramatically since 1970, and a large area of northern California forests has burned (Westerling et al., 2011; Schoennagel et al., 2017). The key factors contributing to increasing wildfires in the western United States are linked to anthropogenic climate change, e.g., increased fuel aridity (Abatzoglou and Williams, 2016), rising temperatures (Schoennagel et al., 2017), and increased drought (Westerling, 2016). Anticipatory climatic models predict a 100% increase in wildfire occurrence in northern California by 2085 (Westerling et al., 2011). Moderate to large increases in wildfires are predicted in wildland-urban interfaces in the next 20 years (Schoennagel et al., 2017; Moritz et al., 2014). Sonoma County is located at the high-risk boundary zone within a wildland-urban interface and has experienced wildfire in the past. However, human settlement

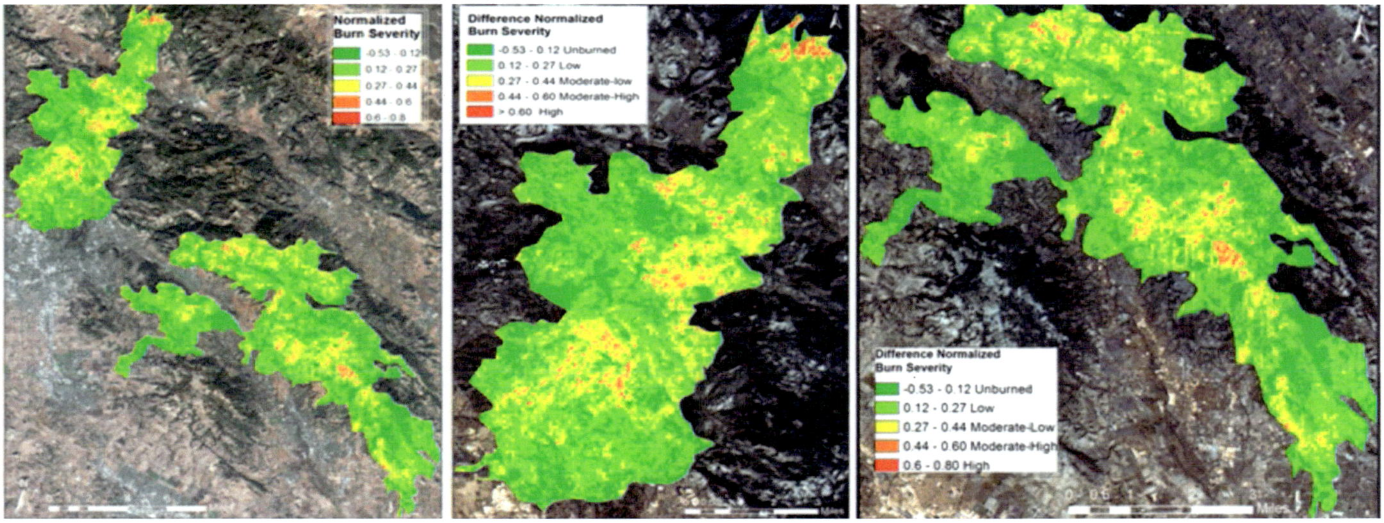

FIGURE 4. The image on the far left depicts the burn severities (e.g. unburned, low, moderate-low, and high burned) for both fires (Tubbs fire at top-left and Nuns fire at bottom-right) north and west of Santa Rosa. The middle image is a higher resolution view at the Tubbs fire and the burn severity distribution. The far right image is the Nuns fire and its associated burn severity results from the analysis. The burn acreage in each severity class was estimated by mult plying number of pixels in each category with pixel size (30 m × 30 m) in a raster calculator. These images serve as visual metrics to burn severity perceptions.

in this area has grown by 20% between 1990 and 2010 (Michels and Sagara, 2018). Despite records of this area's fire risk, human settlement in the area has grown by 20% between 1990 and 2010, indicating that people moving to the area tended to not know about the history of wildfire or were unaffected by the risk. Moreover, housing density in Sonoma County is predicted to grow from 1.81 houses/acre to 3.05 houses/acre between 2010 and 2050, respectively, resulting in future risk to settlements (Mann et al., 2014).

The burn severity maps for both Tubbs and Nuns fires are useful for forest managers to quickly glance at the area to determine priority areas for forest management. Estimation of wildfire size and fire risk zones are important for forest managers to implement thinning practices for fire-prone areas, fuel treatment, such as prescribed burning, and forest health assessment to facilitate recovery and resilience of conifer forests after fire disturbance. Generally, the extent of fire, location, and structural damage depicted in photo galleries are posted on federal web sites. Therefore, our findings are useful to the US Forest Service for the implementation of sustainable forest management practices of vulnerable areas.

This study suggests that of the two methods applied to the burn severity analysis, the method of change detection is superior. Using a pixel-over-pixel iso-clustered supervised classification scheme was more accurate than the dNBR approach. The use of manual classification and raster reclassified image (dNBR) is helpful with heat classifications.

We compared the burn acreage estimated through change detection and dNBR methods with Cal Fire estimation. Our results detected 24% less burned acreage than Cal Fire estimation. However, in terms of accuracy assessment, 75% agreement between Cal Fire and ground control data was obtained. Since we used moderate resolution Landsat data, it impacted accuracy. For example, our accuracy results were comparable with Fisher et al. (2017) land use classification accuracy, in which they obtained 75% for 30 m data and 82% for 1 m data. The fire risk map of Tubbs and Nuns is useful to fire and land managers to look for high-severity sites that are vulnerable to soil erosion and plant regeneration.

CONCLUSIONS & RECOMMENDATIONS—This analysis has drawn the following conclusions for the applied burn severity analysis on Tubbs and Nuns fires in Sonoma and Napa County.

- Change detection determined to be more accurate than the dNBR approach by itself possibly due to the change detection usage of a pixel-over-pixel comparison approach.

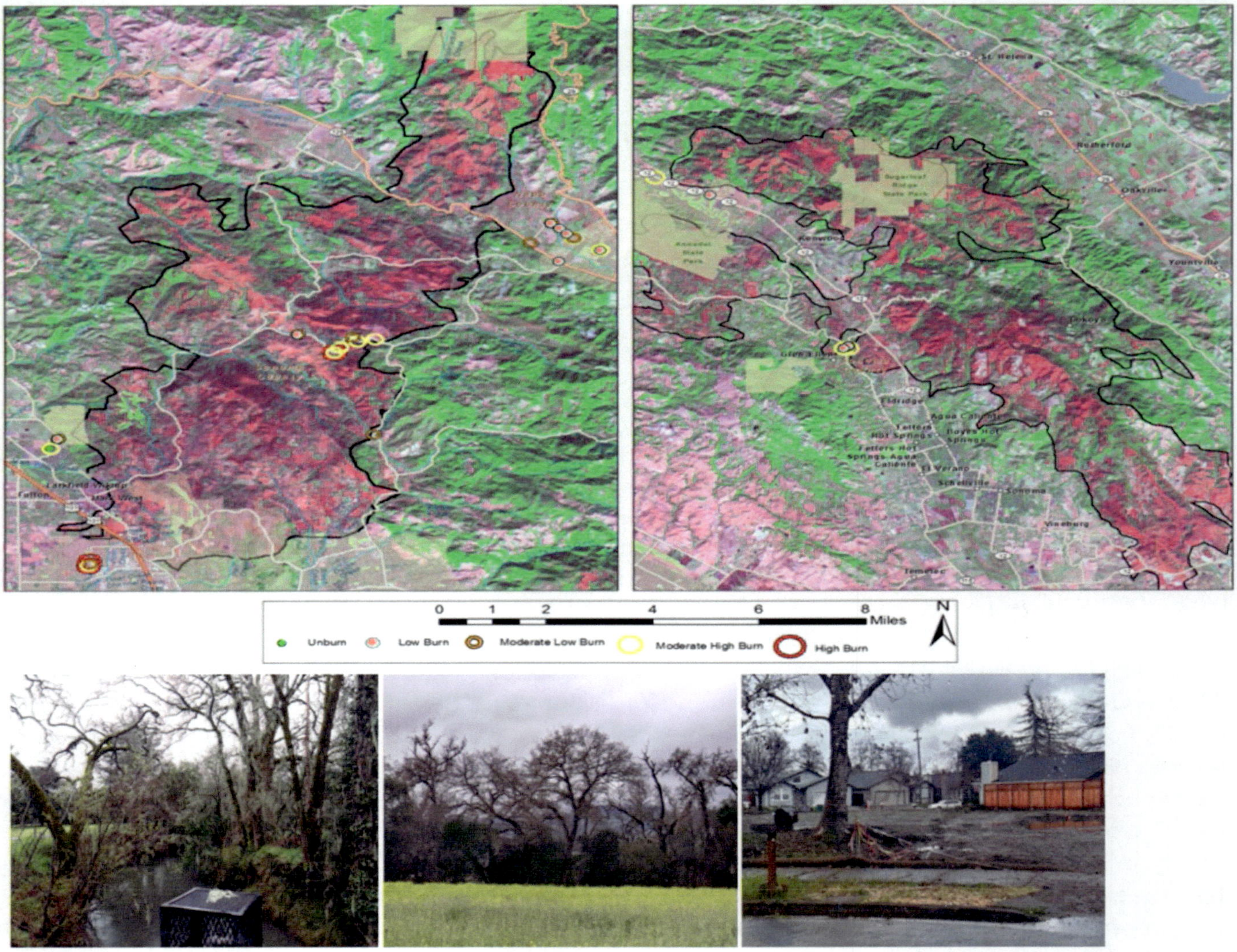

FIGURE 5. A land field survey (ground truthing) using a GPS Trimble Juno Geo 7X to mark the types of burn severities, while comparing a composite map layer of both Tubbs and Nuns fires and ground truthing survey sites. (left to right) Low burn [Geyserville: 38.594037N, -122.597456E], Moderate low burn [Geyserville: 38.710837N, -122.911748E], Moderate high burn [Santa Rosa: 38.472987N, -122.747941E].

- The change detection analysis results identified that the Nuns fire burned an estimated 38,355 acres, while the Tubbs fires burned 25,679 acres.
- A comparison to the actual acreage burned, provided by Cal Fire, indicated that this analysis was on average 76 ± 8% accurate in identifying burn severity on the basis of acreage burned.
- We recommended the use of high resolution images to find post-fire recovery of disturbed forest utilizing the present findings as a baseline study to implement sustainable forest management practices.

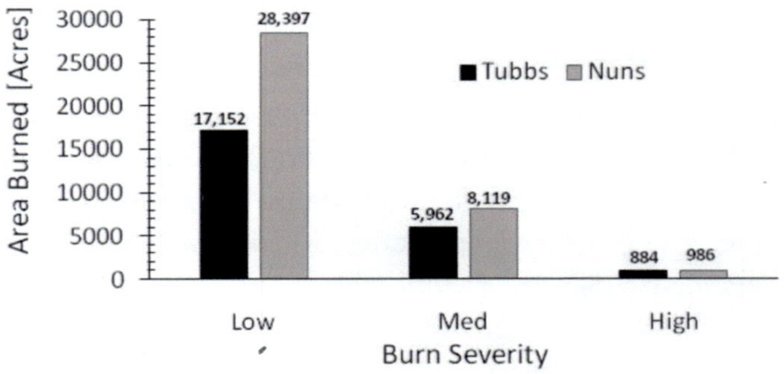

FIGURE 6. Comparison of fire burn severity on an acreage-burned basis for the Tubbs and Nuns fires.

ABOUT THE AUTHORS—John Cortenbach graduated from the Environmental Science Program at HSU in 2018. This paper was based on Cortenbach and his colleague's class project for GSP 326 advised by Dr. Madurapperuma. He is passionate about volunteering for social activities, hiking, and conducting GIS/remote sensing projects. The author is interested in applying remote sensing techniques for real life situations for wildfire mapping and invasive species habitat mapping.

Richard Williams graduated from the Environmental Resources Engineering Program at HSU. He collaborated with Cortenbach for GSP 326 project. Williams is interested in remote sensing application for renewable energy. He was president for the Renewable Energy Student Union (RESU) Club in 2016–2017 and was secretary in 2015–2016.

Dr. Buddhika Madurapperuma is a Lecturer/Research Associate in the Departments of Forestry and Wildland Resources and Environmental Science and Management. He is a major advisor for Cortenbach and William's class project. He teaches GIS, remote sensing, forest ecology, and dendrology classes at HSU. Dr. Madurapperuma conducts multidisciplinary research on remote sensing, i.e., hyperspectral remote sensing for invasive species detection and mapping, ecological studies on invasive species, and post-fire seedling recruitment and seed maturation schedule of CA conifer species.

ACKNOWLEDGMENTS—We would like thank to Carolyn Delevich for making valuable suggestions and edits for the manuscript. We greatly acknowledge reviewers for valuable comments to improve the contents of our manuscript.

REFERENCES

Abatzoglou, J. T., Williams, A. P. 2016. Impact of anthropogenic climate change on wildfire across western US forests. *Proceedings of the National Academy of Sciences of the United States of America* 113: 11770–11775.

Cal Fire. 2017. "California Statewide Maps." CAL FIRE - Home, State of California. Available from:<http://www.fire.ca.gov/current_incidents/> (Dec. 11, 2017).

Exelis. 2015. Exploring ENVI. Chapter 3: Burn severity and ecological risk assessment. pp. Exelis Visual Information Solutions, Inc. 44–60. Available at http://www.harrisgeospatial.com/portals/0/CBT/Mastering_the_Basics_of_ENVI/presentation_content/external_files/Exploring_ENVI52_March2015.pdf (May 6, 2019).

Fisher, J. R., Acosta, E. A., Dennedy-Frank, P. J., Kroeger, T., Boucher, T. M. 2017. Impact of satellite imagery spatial resolution on land use classification accuracy and modeled water quality. *Remote Sensing in Ecology & Conservatopm* 4: 137–149.

Jensen, J. R. 2005. Introductory Digital Image Processing: A Remote Sensing Perspective, 3rd ed.; Prentice-Hall, Inc. Upper Saddle River, NJ, USA.

Madsen, R. S., Haynes, H. J., McCaffrey, S. M. 2018. Wildfire risk reduction in the United States: Leadership staff perceptions of local fire department roles and responsibilities. *International Journal of Disaster Risk Reduction* 27: 451–458.

Mann, M. L., Berck, P., Moritz, M. A., Batllori, E., Baldwin, J. G., Gately, C. K., Cameron, D. R. 2014. Modeling residential development in California from 2000 to 2050: Integrating wildfire risk, wildland and agricultural encroachment. *Land Use Policy* 41: 438–452.

Michels, P., Sagara, E. 2018. Should development be extinguished on California's fire-prone hills? *Journal of Alta California*, July 3, 2018. Available from: https://www.revealnews.org/article/should-development-be-extinguished-on-californias-fire-prone-hills/.

Miller, J. 2017. Where the Fires in Northern California Came From, and What Lies Ahead. The New Yorker, 18 Oct 2017. Available from: www.newyorker.com/tech/elements/where-the-fires-in-northern-california-came-from-and-what-lies-ahead.

Moritz, M. A., Batllori, E., Bradstock, R. A., Gill, A. M., Handmer, J., Hessburg, P. F., Leonard, J., McCaffrey, S., Odion, D. C., Schoennagel, T., Syphard, A. D. 2014. Learning to coexist with wildfire. *Nature* 515: 58–66.

Nor Cal Fires. (n.d.). Google Crisis Map, Google. Available from: <https://google.org/crisismap/google.com/2017-tubbs-fire> (Dec. 8, 2017).

Rozario, P., Madurapperuma, B., Wang, Y. 2018. Remote sensing approach to detect burn severity risk zones in Palo Verde National Park, Costa Rica. *Remote Sensing* 10: 1427.

Schoennagel, T., Balch, J. K., Brenkert-Smith, H., Dennison, P. E., Harvey, B. J., Krawchuk, M. A., Mietkiewicz, N., Morgan, P., Moritz, M. A., Rasker, R., Turner, M. G. 2017. Adapt to more wildfire in western North American forests as climate changes. *Proceedings of the National Academy of Sciences of the United States of America* 114: 4582–4590.

Stewart, I. T., Cayan, D. R., Dettinger, M. D. 2005. Changes toward earlier streamflow timing across western North America. *Journal of Climate* 18: 1136–1155.

USGS. 2017. EarthExplorer. EarthExplorer, U.S. Department of the Interior. Available from: <https://earthexplorer.usgs.gov/> (Nov. 2, 2017).

Watkins, D., Griggs, T., Lee, J. C., Park, H., Singhvi, A., Wallace, T., Ward, J. 2017. How California's Most Destructive Wildfire Spread, Hour by Hour. The New York Times. Available from: <https://www.nytimes.com/interactive/2017/10/21/us/california-fire-damage-map.html> (Dec. 8, 2017).

Westerling, A. L. 2016. Increasing western US forest wildfire activity: Sensitivity to changes in the timing of spring. *Philosophical Transactions of the Royal Academy of Sciences B* 371: 20150178.

Westerling, A. L., Bryant, B. P., Preisler, H. K., Holmes, T. P, Hidalgo , H. G., Das, T., Shrestha, S. R. 2011. Climate change and growth scenarios for California wildfire. *Climatic Change* 109: 445–463.

Westerling, A. L., Hidalgo, H. G., Cayan, D. R., Swetnam, T. W. 2006. Warming and earlier spring increase western US forest wildfire activity. *Science* 313: 940–943.

Williams, A. P., Allen, C. D., Millar, C. I., Swetnam, T. W., Michaelsen, J., Still, C. J., Leavitt, S. W. 2010. Forest responses to increasing aridity and warmth in the southwestern United States. *Proceedings of the National Academy of Sciences of the United States of America* 107: 21289–21294.

SUPPLEMENTARY MATERIAL

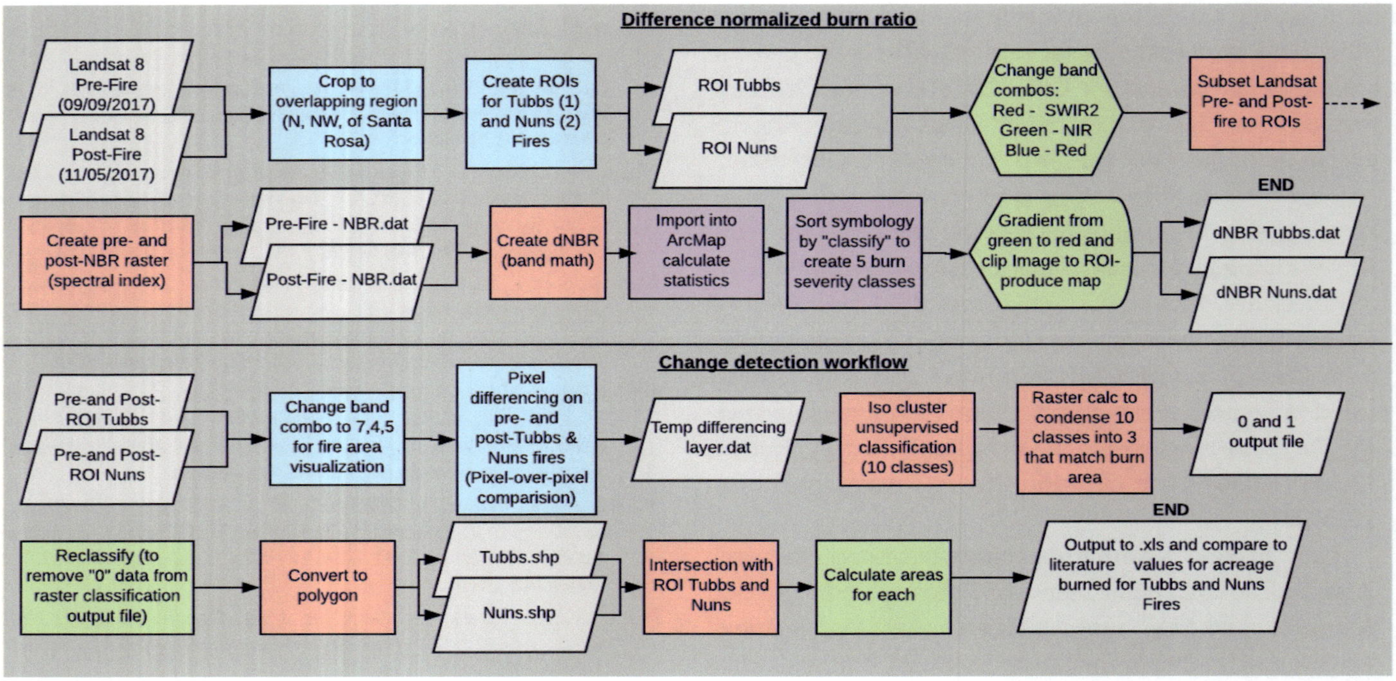

FIGURE S1. Workflow for the two main analyses conducted in order to determine acreage burned and burn severity for both the Tubbs and Nuns fires.

Humboldt County Public Electric Vehicle Charging Station Service Area and Gap Analysis

Chih-Wei Hsu[1*]

KEYWORDS—*electric vehicle, public charging stations, geospatial, service area analysis*

*Corresponding Author: ch104@humboldt.edu
[1]Masters Student
Department of Environmental Systems
Humboldt State University, Arcata, CA 95521

Hsu, C.-W. 2019. Humboldt County public electric vehicle charging station service area and gap analysis. *ideaFest Journal: Interdisciplinary Journal of Creative Works & Research from Humboldt State University* 3: 53–59.

INTRODUCTION—The transportation sector accounts for a third of the total greenhouse gas (GHG) emissions in the United States and 41% of GHG emissions in California (California Air Resources Board, 2018; US Environmental Protection Agency, 2018). The transportation sector GHG emission is greater in rural counties per capita, owing to higher vehicle ownership rates, less comprehensive public transportation systems, and longer trip distances (Federal Highway Administration, 2017; Tamayao, Blackhurst, & Matthews, 2014). In Humboldt County, a mostly rural county, approximately 54% of all GHG emissions came from transportation in 2006 (County of Humboldt, 2012). Addressing high emissions from the transportation sector will play a critical role in achieving California's climate goal of an 80% emissions reduction from 1990 levels by 2050.

Without dramatically changing the conventional auto-centric view of mobility in the United States, in northern California the most cost-effective near-term way to reduce the transportation sector GHG emissions is by electrifying the vehicle fleet (Fingerman, 2018). California has been a leader in promoting Zero Emissions Vehicles (ZEV) and the recent executive order B-48-18, Zero Emissions Vehicle Executive Order, committed California to adopt five million ZEVs by 2030. There are approximately 30.6 million vehicles registered in California; of that, 26.9 million vehicles are in non-rural counties (California Department of Motor Vehicles, 2018a). It is possible for California to achieve the five million ZEV goal in non-rural counties alone, however, the ZEV adoption should be equitable, with benefits spreading across demographic and geographic locations. Furthermore, under policies such as California's Low Carbon Fuel Standard, a lopsided ZEV adoption toward urban areas would result in rural areas subsidizing the urban low carbon fuel uptake (Fingerman, 2018). Beyond the GHG reduction benefits, electric vehicle (EV) adoption creates societal benefits. Direct benefits include reduced transportation expenses, such as fuel and maintenance costs. Other less understood benefits include local economic development and grid stabilization potential (Malmgren, 2016).

To achieve an equitable EV adoption across demographic groups, price barrier and range barrier (more prevalent in used EVs), among others, would need to be overcome. Although financial incentive programs such as Clean Vehicle Rebate Project and Federal Tax Credit help lower the purchasing cost of electric vehicles, the post-incentive price of a new EV is still prohibitive to low-income households. This is evident through the Clean Vehicle Rebate Project survey. Of the survey respondents, 95% of EV owners have an annual household income higher than $50,000 and 76% of EV owners have an annual household income higher than $100,000 (Center of Sustainably Energy, 2017). Another study shows new EV owners have an average household income of $227,000 and used EV owners had an average household income of $173,400 (Turrentine et al., 2018). There are few programs such as the Electric Vehicle Assistance Program that are aimed at increasing EV adoption in low-income families, but no comprehensive plan has been developed both statewide and regionally in the North Coast. If policies were to be designed to increase EV adoption in rural low-income families, lower cost used EVs are likely to be the focus. However, the nascent used EV market does not offer many vehicle options currently. A first generation used EV has a lower price barrier but comes with limited range (e.g. used 2011–2016 Nissan Leaf with 50 to 70 miles of range [Plug-in America, n.d.]) thus creating a higher range barrier.

One way to lower the range barrier is to have a reliable and expansive public charging network. This study investigates the current public charging network comprehensiveness in Humboldt County by analyzing the public EV charging station service area coverage using the Network Analyst Tool in ArcMap. The result of the analysis will provide both visual and quantitative representations of the public EV charging network and identify the coverage gap for relatively high building density areas in Humboldt County.

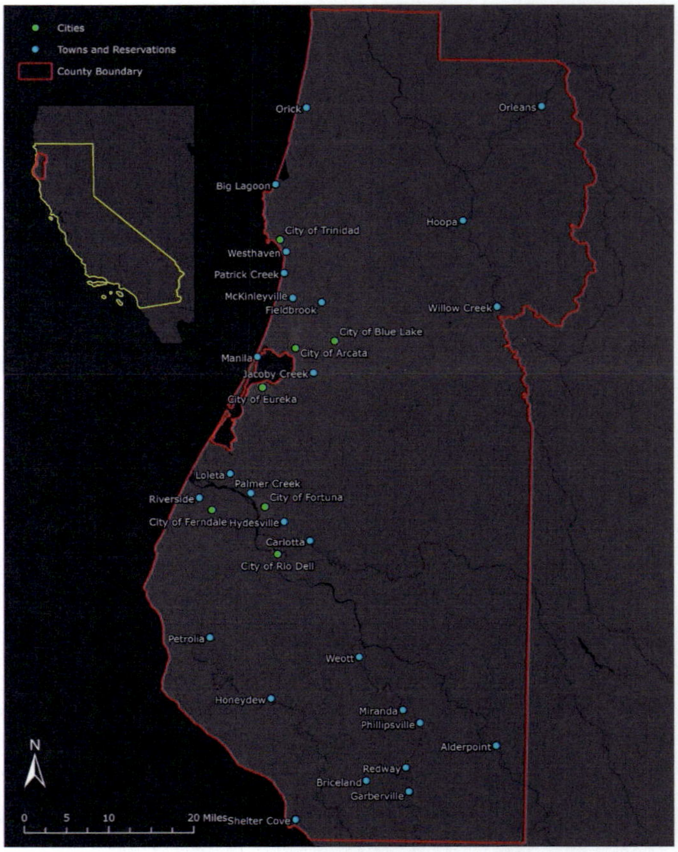

FIGURE 1. Humboldt County Map.

METHODS—*Data sources*. Geospatial data of Humboldt County used for the project includes Microsoft building footprint from GitHub (open source), the road way centerline, city, community service district, and county boundary shapefiles from Humboldt County Geospatial Information System Department website, digital elevation model raster files from US Geological Survey's The National Map website, public electric vehicle charging station locations obtained from Department of Energy's Alternative Fuels Data Center, and the 2017 Local Area Transportation Characteristics for Households survey data from Bureau of Transportation Statistics.

Service area analysis. Service area analysis on Humboldt County charging stations was done in ArcMap with the Network Analyst toolset extension. There were two types of service areas, one based on the traveling time and one based on traveling distance. The following steps were used to create the Humboldt County public EV charging station network file. First, Z (elevation) information was added to the road way centerline polyline shapefile with the mosaiced digital elevation model raster file using the Add Surface Information tool. Then, in the attribute table, the travel time for each road section was calculated by dividing the road section length by the speed limit. Finally, network file was created using the road way centerline file with the road length and travel time as cost attributes.

To perform the service area analysis, a new service area layer was created and the EV charging stations point shapefile was added as "facilities." For the travel time service area analysis, the default breaks in the analysis setting were set to 5, 15, 30, and 60 minutes. For the travel distance service area analysis, the default breaks in the analysis setting were set to 5, 15, 30, and 60 miles. For both analyses, travel direction was set as "towards facility." Finally, service area polygons buffered (called "Trim Polygons" in analysis setting) to 0.25 miles from the roads were generated for each analysis.

The portions of total buildings in the county covered by the current EV charging station service areas—both travel time and travel distance service areas—were calculated. This was done by dividing the number of buildings located within a given distance (i.e., 5, 15, 30, or 60 mi) or within a given traveling time (i.e., 5, 15, 30, or 60 min) from all EV charging stations by: a) the number of buildings in each city or town and b) the total number of buildings in the county.

Building density analysis. Microsoft building footprint GeoJSON file was read in as polygon shapefile first and then converted to point shapefile. To analyze the building density, the point shapefile was converted into a raster file using Point Density tool with building density per square-mile as the variable. The settings used for the tool were the following: "NONE" for the population field, 337.19 meters for the cell size setting, circle for the neighborhood setting, and 907.78 meters for the radius setting (to calculate the density per square mile).

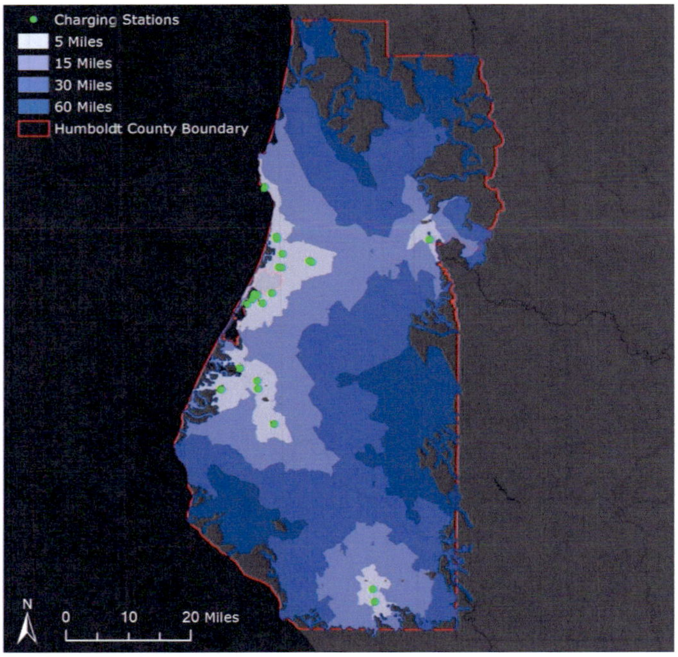

FIGURE 2. Public EV charging station service areas in Humboldt County based on traveling distance to each station. The grey areas within the county boundary indicate areas beyond the 60-mile service coverage.

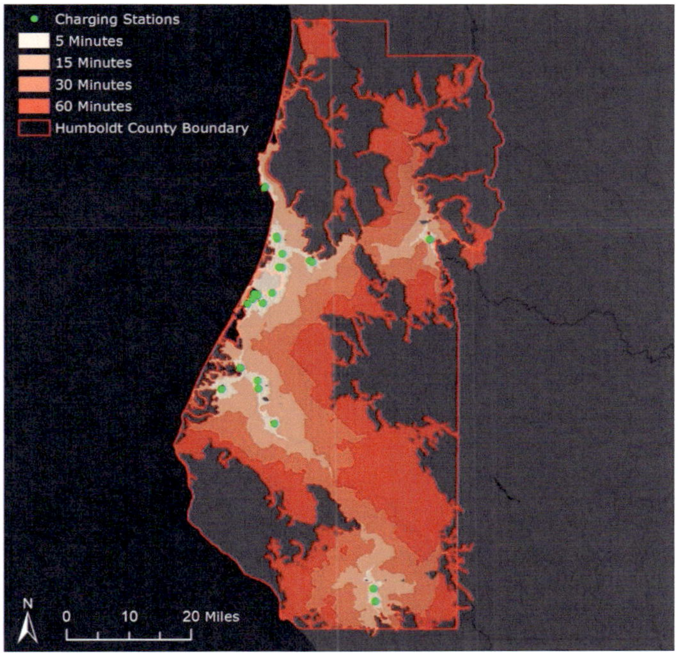

FIGURE 3. Public EV charging station service areas in Humboldt County based on travel time to each station. The grey area within the county indicates areas beyond the 60-minute service coverage

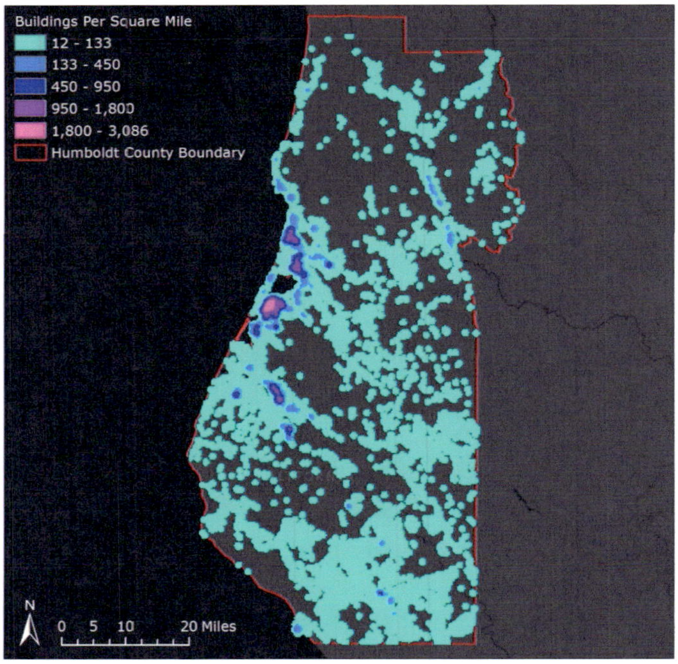

FIGURE 4. Building density map of Humboldt County. Grey shading indicates areas with less than 12 buildings per square mile.

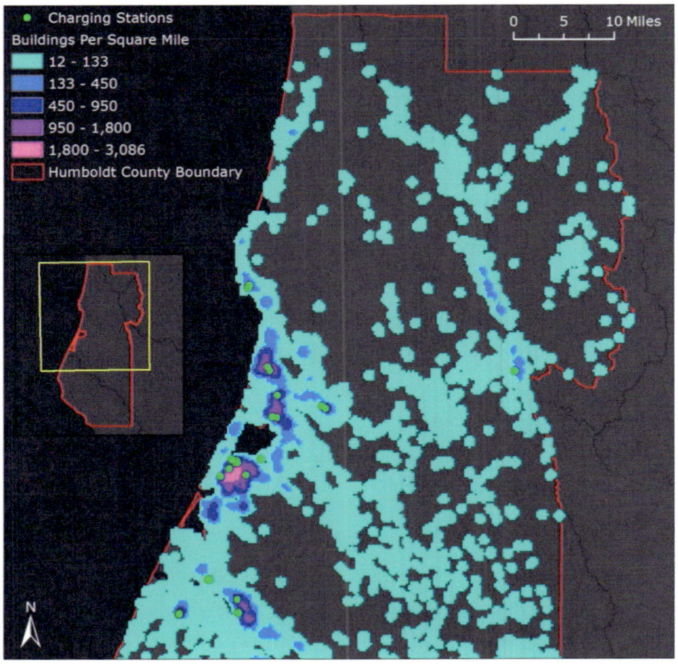

FIGURE 5. Public EV charging station locations overlaid with building density for northern Humboldt County. Areas with less than 12 buildings per square mile are in grey.

RESULTS—Humboldt County's 31 public EV charging station locations are located mainly in McKinleyville, the Humboldt Bay (Eureka-Arcata) area, and towns around Eel River (i.e., Loleta, Ferndale, and Fortuna) with the exception of Willow Creek, Blue Lake, Trinidad, Rio Dell, Garberville, and Benbow (FIG 1 & 2).

Service coverage by travel distance. At the five-mile service area level, service coverage (in terms of percent buildings covered) reaches 76%—approximately three in four buildings—county wide (TABLE 1). At the thirty-mile service area level, the EV charging station service network is connected across the county (FIG 2) and service coverage reaches 96% county wide (TABLE 1).

Service coverage by traveling time. The five-minute EV public charging service area level covered 62% of buildings in the county. At the thirty-minute EV public charging service area level, 91% of buildings in the county are covered and the service areas are connected across the county (FIG 3 & TABLE 1).

Service coverage and building density. Public EV charging stations in the county are located in areas with relatively high building density, however, not all areas with relatively high building density have public EV charging stations. Building density is highest in Humboldt Bay area (Eureka-Arcata), McKinleyville, and Fortuna (FIG 4). In addition to the aforementioned areas, public EV charging stations are also found in Trinidad, Blue Lake, Willow Creek, Ferndale, Rio Dell, Loleta, Garberville, and Benbow, just south of Garberville (FIG 5 & 6).

Service coverage gaps. For the purpose of the analysis, a high building density area is defined as areas with greater than 200 buildings per square mile. In the county, Hoopa, Miranda, and Shelter Cove are the only towns with high building density areas that are not covered by the five-minute and five-mile level public EV charging service area (FIG 7–10). Shelter Cove is the only town with high building density areas that are not covered by the fifteen-minute level public EV charging service area (FIG 8 & 10).

DISCUSSION—Based on the service area analysis result, more than three quarters, 76%, of the buildings in Humboldt County are within a five-mile distance of a public charging station and 62% within a five-minute travel time of public charging stations. This suggests that the public EV charging station coverage in Humboldt County is serving a majority of the buildings and population, however, not without some areas left behind. The

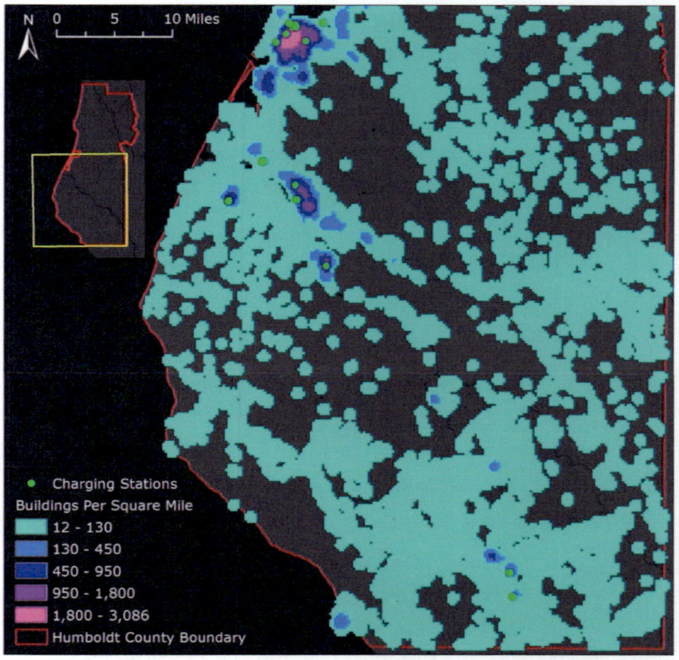

FIGURE 6. Public EV charging station locations overlaid with building density for southern Humboldt County. Grey shading indicates areas with less than 12 buildings per square mile.

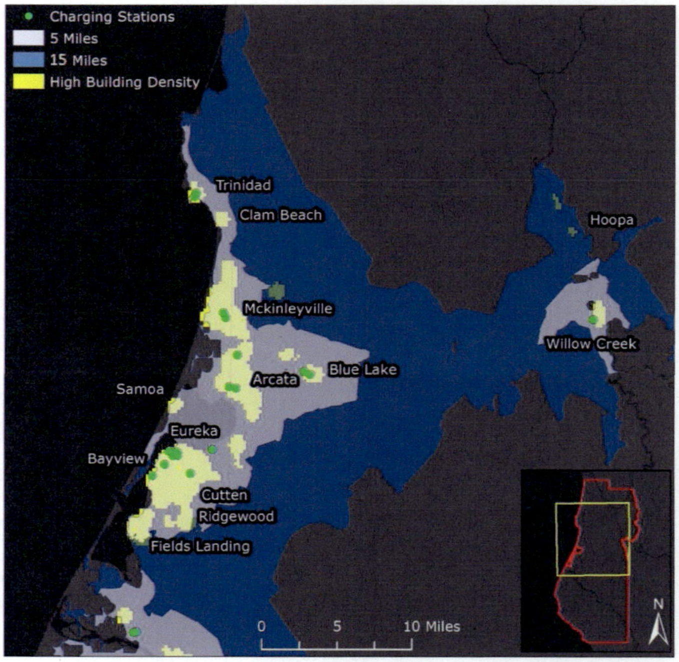

FIGURE 7. Northern Humboldt County high building density areas and public EV charging station coverage at five- and fifteen-mile level. Grey shading indicates areas beyond fifteen-mile service coverage.

Hoopa reservation, Miranda, and Shelter Cover are three high building density areas outside of the five-minute and five-mile EV public charging service coverages. Furthermore, there is currently no EVs registered in Hoopa (California Department of Motor Vehicle, 2018b). As EV market penetration increases, areas such as Hoopa will likely need public EV charging stations locally to support the local EV adoption.

A first generation used EV (FG-EV) has an approximate 60-mile range. Based on the 30- and 60-mile service area result, it is possible, however not necessarily practical, for the FG-EV to travel throughout the county. The 30-mile service area map (FIG 2) shows the round-trip limit for the FG-EV. To make longer travel by FG-EV

TABLE 1. Public EV charging station service area coverage, in terms of buildings covered, for cities and towns in Humboldt County based on distance and travel time. Note county-wide summary includes buildings located in other unincorporated areas (i.e. not in the following cities and towns). CSD stands for community service district.

Name	Total Buildings	Five Mi (%)	Five Mins (%)	Fifteen Mi (%)	Fifteen Mins (%)	Thirty Mi (%)	Thirty Mins (%)
Alderpoint	194	0	0	0	0	100	50
Arcata	27,030	100	23	100	49	100	74
Big Lagoon	334	0	0	97	32	100	66
Blue Lake	3,218	100	25	100	50	100	75
Briceland	123	0	0	100	33	100	67
Carlotta	1,395	0	0	100	26	100	63
Eureka	72,717	100	21	100	48	100	74
Ferndale	3,294	99	25	99	50	99	75
Fieldbrook	3,416	46	8	100	38	100	69
Fortuna	21,869	100	24	100	49	100	75
Garberville	1,994	100	21	100	48	100	74
Hoopa	2,912	0	0	87	19	100	59
Humboldt CSD	31,387	96	16	100	44	100	72
Hydesville	2,328	32	0	99	33	99	66
Jacoby Creek	1,973	90	7	100	38	100	69
Loleta	1,248	100	25	100	50	100	75
Manila	1,527	95	5	100	36	100	68
McKinleyville	27,034	99	23	100	49	100	74
Miranda	566	0	0	100	31	100	66
Orick	763	0	0	19	4	100	51
Orleans	264	0	0	0	0	0	0
Palmer Creek	668	100	25	100	50	100	75
Patrick Creek	54	100	0	100	33	100	67
Phillipsville	291	0	0	100	33	100	67
Redway	2,296	97	14	98	43	98	71
Rio Dell	6,045	100	25	100	50	100	75
Riverside	644	89	19	99	46	100	73
Trinidad	1,344	100	25	100	50	100	75
Weott	336	0	0	0	0	100	50
Westhaven	1,267	100	25	100	50	100	75
Willow Creek	3,912	98	19	100	46	100	73
County Wide	**68,405**	**76%**	**62%**	**88%**	**83%**	**96%**	**91%**

more practical, more public direct current fast chargers would need to be installed, however, not all FG-EV are compatible with fast chargers. The new model EVs provide more than 200 miles from a single full charge; thus range barrier does not post an issue for traveling within the county.

Although there is a temporal aspect—service areas based on traveling time—of this study, it does not investigate the charging time nor the delay in travel time based on charging station capacity (i.e. available plugs versus EV charging demand). To better understand the adequacy of the county's public charging infrastructure in this regard, EV travel modeling would need to be performed. An example of electric vehicle modeling has been done by the Schatz Energy Research Center (Schatz Energy Research Center, 2013).

The analytical method of this study can be applied in future studies to shed light on the relationship between EV adoption and charging network coverage to better understand the driving factors for EV adoptions.

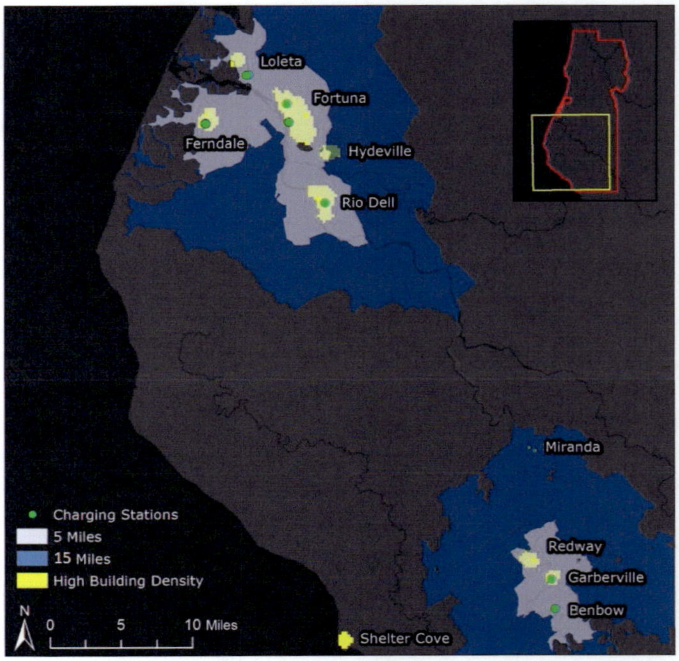

FIGURE 8. Southern Humboldt County high building density areas and public EV charging station coverage at five- and fifteen-mile level. Grey shading indicates areas beyond fifteen-mile service coverage.

CONCLUSION—EV adoption has multiple barriers;

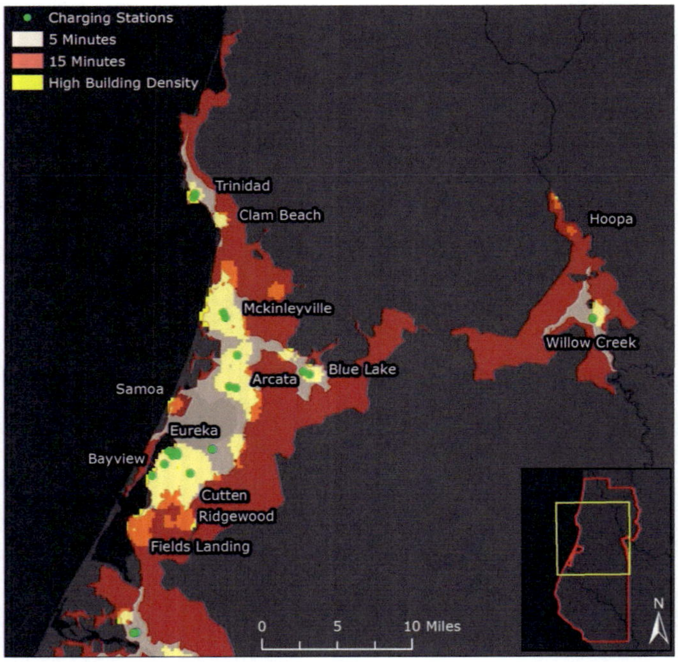

FIGURE 9. Northern Humboldt County high building density areas and public EV charging station coverage at five- and fifteen-minute level. Grey shading indicates areas beyond fifteen-minute service coverage.

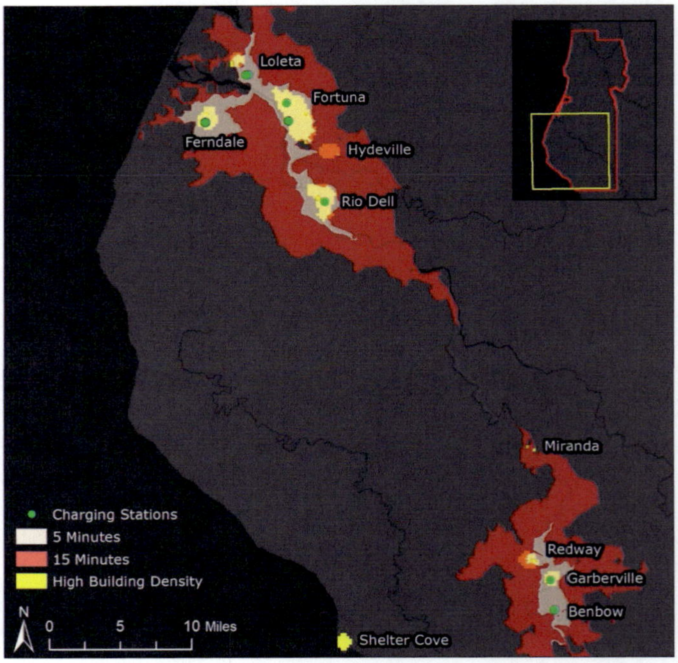

FIGURE 10. Southern Humboldt County high building density areas and public EV charging station coverage at five- and fifteen-minute level. Grey shading indicates areas beyond fifteen-minute service coverage.

this study focused on the range barrier by investigating the public EV charging station coverage in Humboldt County. The study found over 60% of buildings in the county are either within five minutes or five miles of travel from a public charging station. The study also found that Hoopa, Miranda, and Shelter Cove are the only areas with more than 200 buildings per square mile that are not currently within five minutes or five miles of travel from a public charging station. As EV market penetration inevitably increases, there may be a need to expand the public charging network to these areas.

ACKNOWLEDGMENTS—The author is grateful for the valuable inputs from thesis advisor Dr. Kevin Fingerman, thesis committee member Nick Lam, Geospatial professor Dr. Jim Graham, Schatz Research Lab researcher Jerome Carman, and fellow EV advocate Derek Ichiena for data sources, analytical methods, and ArcMap© troubleshooting.

REFERENCES

California Air Resources Board. 2018. California greenhouse gas emissions for 2000 to 2016: Trends of emissions and other Indicators. Retrieved from https://www.arb.ca.gov/cc/inventory/pubs/reports/2000_2016/ghg_inventory_trends_00-16.pdf.

California Department of Motor Vehicles. 2018a. California motor vehicle fuel types by county [PDF]. Retrieved from https://www.dmv.ca.gov/portal/dmv/detail/pubs/media_center/statistics.

California Department of Motor Vehicles. 2018b. California motor vehicle fuel types by ZIP code [PDF]. Retrieved from https://www.dmv.ca.gov/portal/dmv/detail/pubs/media_center/statistics.

Center for Sustainable Energy. 2017. Summary documentation of the electric vehicle consumer survey, 2013–2015 edition. Retrieved from https://cleanvehiclerebate.org/sites/default/files/attachments/CVtConsumerSurvey2013-15Reference.pdf.

County of Humboldt. 2012. Climate action plan: A strategy for greenhouse gas reduction and adaptation to global climate change. Retrieved from https://humboldtgov.org/DocumentCenter/View/1347/Draft-Climate-Action-Plan-PDF?bidId=.

County of Humboldt. 2017. Humboldt county shapefile [Zipped shapefile]. Retrieved from https://humboldtgov.org/276/GIS-Data-Download.

County of Humboldt. 2013. Roadway centerline shapefile [Zipped shapefile]. Retrieved from https://humboldtgov.org/276/GIS-Data-Download.

Federal Highway Administration. 2017. Planning for transportation in rural areas. Retrieved from https://www.fhwa.dot.gov/planning/publications/rural_areas_planning/page03.cfm.

Fingerman, K. R., Sheppard, C., & Harris, A. 2018. California's low carbon fuel standard: Modeling financial least-cost pathways to compliance in northwest California. *Transportation Research Part D: Transport and Environment* 63: 320–332.

Malmgren, I. 2016. Quantifying the societal benefits of electric vehicles. *World Electric Vehicle Journal* 8: 996–1007.

Microsoft. 2018. US building footprints: California [Zipped geojson]. Retrieved from https://github.com/Microsoft/USBuildingFootprints.

Plug-in America, n.d. Used electric car buyers' guide. Retrieved from https://pluginamerica.org/buying-a-used-electric-car/.

Schatz Energy Research Center. 2013. Macro-scale siting of plug-in electric vehicle infrastructure for the North Coast of California. Retrieved from http://www.schatzlab.org/projects/policyanalysis/pev/.

Tamayao, M. M., Blackhurst, M. F., & Matthews, H. S. 2014. Do US metropolitan core counties have lower scope 1 and 2 CO_2 emissions than less urbanized counties? *Environmental Research Letters* 9: 104011.

Turrentine, T., Tal, G., Rapson, D. 2018. The dynamics of plug-in electric vehicles in the secondary market and their implications for vehicle demand, durability, and emissions: A research report from the National Center for Sustainable Transportation. Retrieved from https://ncst.ucdavis.edu/wp-content/uploads/2014/08/NCST-ARB_Turrentine-Tal-Rapson-Dynamic-of-PEVs-in-Secondary-Market_Final-Report_APR-2018.pdf.

US Department of Energy. 2018. Alternative fuel stations: California [CSV]. Retrieved from https://afdc.energy.gov/data_download/.

US Department of Transportation. 2017. Local area transportation characteristics for households survey data [CSV]. Retrieved from https://www.bts.gov/latch/latch-data.

US Environmental Protection Agency. 2018. Sources of greenhouse gas emissions. https://www.epa.gov/ghgemissions/sources-greenhouse-gas-emissions.

US Geological Survey. 2013. Elevation products (3DEP): 1/3 arc-second DEM: Humboldt County [Zipped raster]. Retrieved from https://viewer.nationalmap.gov/basic/.

Identification of Mined Areas That May Contribute to Water Quality Degradation at Hobet Coal Mine, West Virginia

Brian P. Murphy[1*]

ABSTRACT—Operational from 1974–2015, the Hobet Coal Mine is one of the largest surface coal mines in the United States. Mining has impacted nearly every aspect of the geography of this region, especially the streams and rivers that are now contaminated with heavy metals, including selenium and iron. Using an erosion model, I identified specific regions that may be subject to high erosion risk in the Rich Hollow basin of the Hobet Coal Mine in West Virginia. This basin is a 444-acre watershed located on the eastern side of the mining complex, and tests positive for high amounts of iron. Data on stream contamination were correlated with the sub-basins that drain into the contaminated regions of the rivers and stream. The erosion model allowed for the identification of areas of particular basins that may have a higher chance of contributing heavy metal contaminants. While full remediation will not be possible, geospatial analysis can be utilized to develop strategies that can assist in the mitigation of some of the worst long-term effects of this type of mining.

KEYWORDS—*GIS, coal, erosion, RUSLE, iron, Hobet Mine, mountaintop removal, valley fill, West Virginia, overland flow*

INTRODUCTION—The Hobet Coal Mine in West Virginia was an active mining site for 41 years and one of the largest surface coal mines in the United States (FIG 1). The impacts of this particular mine are especially pronounced because of the sheer scale of alteration to the landscape caused by mountaintop removal (MTR) to retrieve the coal and valley fill (VF) mining techniques to dispose of the mining debris. In order to reach the coal located in shallow seams below these Central Appalachian Mountain peaks, up to 650 vertical feet of earth was demolished (Bernhardt and Palmer 2011). Explosives and massive machines were used, such as the 20-story tall dragline excavator, which can weigh up to eight million pounds and remove up to 110 cubic yards of earth with one scoop of its bucket (Fox 1999). The earth that did not contain coal was packed into the adjacent valleys, creating valley fills up to one mile long and 1000 feet wide (Hendryx and Holland 2016). What remains in the wake of this coal extraction is a scarred landscape consisting of artificially flattened plateaus devoid of the forests that once covered the land.

*Corresponding Author: bpm213@humboldt.edu
[1]Alumnus
Department of Environmental Science and Management
Humboldt State University, Arcata, CA 95521

Murphy, B. P. 2019. Identification of mined areas that may contribute to water quality degradation at Hobet Coal Mine, West Virginia. *ideaFest Journal: Interdisciplinary Journal of Creative Works & Research from Humboldt State University* 3: 60–65.

This mining has impacted nearly every aspect of the geography of the region. The soil has been compacted from heavy machinery, reducing permeability and increasing the overland flow of water (Griffith et. al 2012). Compacted soils have contributed to high mortality rates of native hardwood trees that are planted in an attempt to remediate the area. The hydrologic system of the mined area is radically altered, especially by the creation of VFs. In particular, the creation of VFs buries the headwaters of the watershed, influencing downstream biotic and chemical conditions. These streams are critical for transporting tree litter downstream, which drives the aquatic food web (Bernhardt and Palmer 2011). Many toxic chemicals

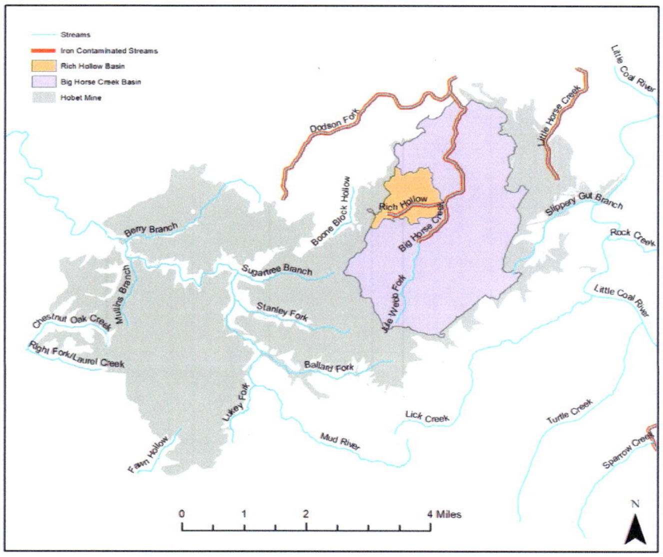

FIGURE 1. Location of the Rich Hollow basin in relation to Big Horse Creek basin. These basins are comprised of areas that were mined in the 1990s.

frequently found alongside coal are released by the mining. Due to oxidation in the presence of water and air, some minerals become new compounds that have higher toxicity. These compounds accumulate in the water and in living things ("bioaccumulate") in the food web where they can cause population collapses of native fish species (Arnold 2014). Some compounds such as pyrite oxidize to generate sulfuric acid and iron hydroxides that drastically lower the pH of the water, sometimes below the threshold of what aquatic communities can withstand. Acids also lead to rapid weathering of rocks and soil which contributes to additional metals into the water system. Another major heavy metal contaminant present at the Hobet mine is selenium, frequently found in and around coal streams. Although it is an essential macronutrient, selenium is very toxic in higher amounts and can bioaccumulate in the local community, resulting in many issues including developmental deformities in fish that can be lethal and potentially cause population collapse (Arnold 2014). These changes caused by mountaintop removal put the Central Appalachian Mountain ecoregion in peril. This biodiversity hotspot is considered to be the most biologically diverse freshwater ecosystem in North America and contains 10% of global salamander and freshwater mussel diversity (Bernhardt and Palmer 2011). Not only do these practices devastate the flora and fauna of the region, but they also inflict an enormous negative impact on humans living in the surrounding communities. In 1999, approximately 450,000 West Virginians (25% of the population) were without drinkable water due to contamination and disruption of aquifers and wells from mining (Fox 1999).

According to data obtained from the West Virginia Department of Environmental Protection, the streams and rivers surrounding the Hobet mine site are contaminated with a variety of heavy metals, most notably selenium and iron. Heavy metal contaminants are usually introduced into the water system by sediment particles (Hudson-Edwards 2003; Yenilmez et. al 2011). The main driver of sedimentation of water systems is overland flow, where water travels over the surface of the land, detaching soil particles from the parent material and transporting them into the stream (Liu et al. 2012). Models that factor in the highest possible rates of erosion can reveal areas that are contributing the most sediment into the stream system. Although areas at high risk for erosion cannot be assumed to be the only contributors of heavy metals into the stream system, they are a good place to conduct further analysis.

Annual rates of erosion can be determined by the use of models such as physical models, based on recreating exact mathematical conditions, and empirical models, based on observations and recorded data (Demirci and Karaburun 2011). Empirical models for estimating soil erosion have less stringent data requirements than physical models, making them more attractive for modeling sites that have less data. The Revised Universal Soil Loss Equation (RUSLE) method is the most commonly used soil erosion model in the world because it can provide estimates of the spatial distribution of soil loss. It does so by analyzing land use, conservation practices, soil type, precipitation, and topography to estimate the effects of precipitation and overland flow on erosion (Demirci and Karaburun 2011).

One area in particular that is a good place to start analyzing the Hobet Mine is the Rich Hollow basin, located in the northeast portion of the Hobet Mine Complex (FIG 1). This hollow is contaminated with high levels of iron, and feeds into the larger stream, Big Horse Creek, which also has tested positive for iron contamination (West Virginia Department of Environmental Protection 2018). This area was mined in the 1990s but the ground still continues to leach heavy metals into the stream system. The damage to the environment inflicted by the exploitive practice of MTR and VF is so dramatic that it is not reversible. By removing forest habitat, destroying the landscape, and polluting the streams, MTR is contributing to the defacing of Earth and to the devastation of local biological and human communities. The legacy of MTR will remain in these communities for generations and will affect the environment for millennia (Lechner et al. 2017). While full remediation will not be possible, geospatial analysis can be utilized to develop strategies that can assist in the mitigation of some of the worst long-term effects of this type of mining. Analyzing the amount of erosion can help identify areas that are more prone to contributing contaminants into the water system.

METHODS—*Hydrology*. This analysis was performed in ArcMap© 10.5.1. Several types of raster and vector data were utilized. The majority of these data was obtained from the West Virginia Department of Environmental Protection (WVDEP) data download, including LiDAR, polyline, and polygon data (West Virginia Department of Environmental Protection 2018). National Agriculture Imagery Program (NAIP) imagery from the United States Department of Agriculture (USDA) was used for land use classification that identified mined areas. The vector data obtained from WVDEP included polyline data identifying the stream network, as well as what types of heavy metals were contaminating each stream, and polygon data of valley fill locations and areas permitted for mining. First, a digital elevation model (DEM) was created from a LiDAR point cloud. This DEM was resampled from 0.5 meter resolution to 6 meter resolution in order to reduce the computational resources required. Next, a hydrologic model was created based on the DEM. The components of the hydrologic model included a DEM that has had its sinks filled, a flow accumulation raster, a flow direction raster, and a stream network raster as well as a stream polyline network. From the flow direction raster, sub-basins were identified. The streams that were found to contain dangerous levels of contaminants were corresponded with the higher resolution stream polyline output from the hydrologic model. From here the sections of the rivers that had been contaminated were corresponded with the basin that drains into it. This was done by creating a pour point at the pixel of the flow accumulation raster right before the stream in question drained into a higher order stream. A watershed function was run to identify all of the streams that flow into the pour point, creating a basin. The resulting raster was then converted to a polygon.

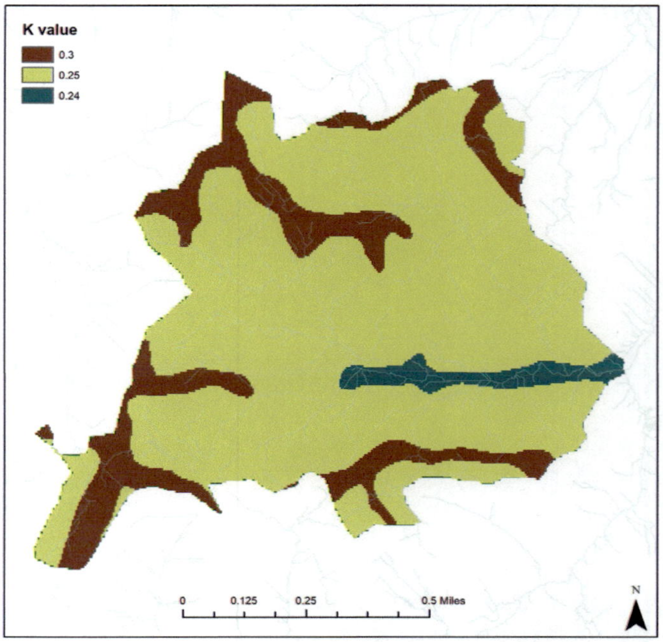

FIGURE 2. Soil erodibility factor (K) of the Rich Hollow basin. Larger values (brown) indicate areas that are comprised of soils that are more susceptible to erosion and small values (blue) indicate areas that are less susceptible to erosion

Erosion. Upon the identification of problematic basins, erosion modeling was performed in order to detect which subregions of the problematic basins were at greatest risk of erosion and were contributing contaminated soil to the stream system. The RUSLE model was used to estimate the spatial distribution of soil erosion at the Hobet Coal Mine. This equation is:

$$A = R \times K \times LS \times C \times P \quad (1)$$

A is the estimated average annual soil loss in tons per acre per year; *R* is the rainfall and runoff erosivity factor; *K* is the soil erodibility factor; *L* is the slope length factor; *S* is the slope steepness factor; *C* is the land cover and land management factor; *P* is the support practice factor. Each factor was calculated using the raster calculator. To determine the *R* (rainfall and runoff erosivity) factor for 2011, when the LiDAR measurements used to create the DEM were taken, the EPA's Rainfall Erosivity Factor Calculator for Small Construction Sites was used (EPA 2018). The *K* (soil erodibility) factor was determined based on

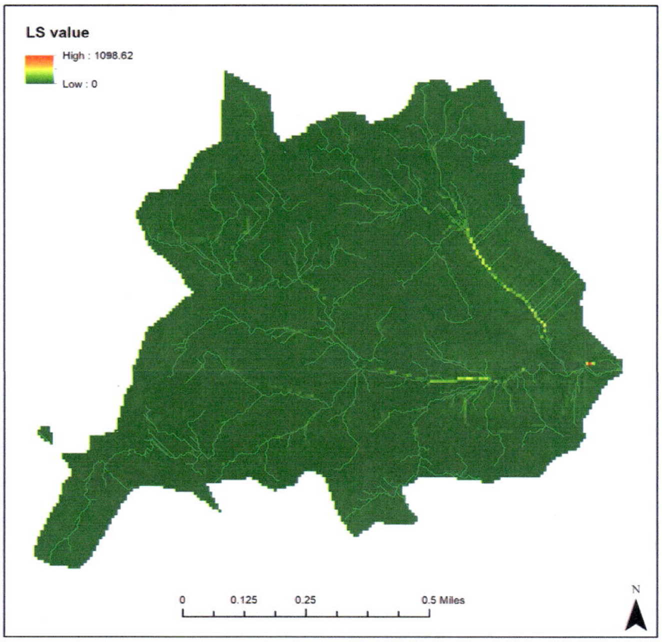

FIGURE 3. The product of slope length (*L*) and slope steepness (*S*) factor values of the Rich Hollow Basin. Larger values (yellow and red) indicate terrain that is more susceptible to erosion.

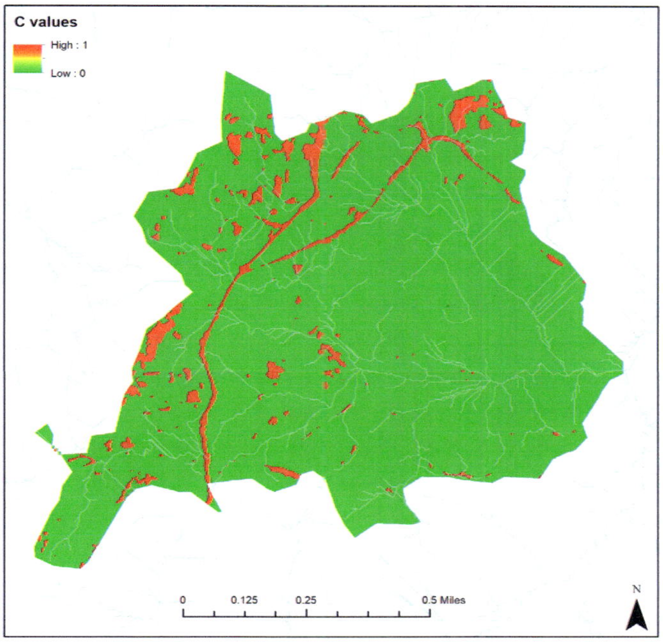

FIGURE 4. The land cover (*C*) factor of the Rich Hollow basin. Aerial imagery was classified into forest, grassland, and exposed soil. High values (red) indicate exposed soil that is significantly more susceptible to erosion.

literature values and published soil surveys including the USDA's Web Soil Survey (USDA 2018; FIG 2). The slope length (*L*) and slope steepness (*S*) factors were calculated using the formula:

LS = Power("richhollow_flowaccumulation.tif"*CellSize/22.1, 0.4) * Power (Sin("richhollow_slope.tif"*0.01745)/0.09, 1.4)*1.4 (2)

Flow accumulation is the raster layer calculated in the hydrology component of this paper, cell size is the spatial resolution of the DEM, and the slope is a raster image calculated by running the slope tool on the DEM (FIG 3). The *C* (land cover and land management) factor was determined by running a supervised classification to identify land cover, with a particular focus on identifying exposed soil, grassland, and forest (FIG 4). These classes were reclassified according to literature values (Kim 2014). Finally, the *P* factor (support practice) was estimated to be one because there were no agricultural erosion suppression practices being used at this site (Demirci and Karaburun 2011).

The resulting raster images for each factor were multiplied together using the raster calculator, resulting in a raster image that depicts the spatial distribution of erosion potential in the units of tons of soil lost per hectare per year. This raster was then classified based on a previous study using the RUSLE method conducted by Demirci and Karaburun (2012) on the severity of erosion potential with 1 t/ha/yr classified as low erosion and 10 t/ha/yr classified as severe erosion. The low erosion rate is classified as 1 t/ha/yr because any greater erosion rate would take 50–100 yr to reverse (Kouli et. al 2008). These results were then depicted on a map, with a spatial resolution of 6 m, which shows areas of high erosion potential with the stream layers calculated from the hydrological model. Zonal statistics were then calculated to find the mean erosion rate of the total area as well as the valley fills.

RESULTS—The RUSLE model was able to generate

TABLE 1. Comparison of erosion rates on valley fills versus the entire area

Site	Erosion Rate (t/ha/yr)	Area (acres)
Total Rich Hollow Basin	7.48 ± 67.28	444.29
Just Valley Fills	5.22 ± 56.17	241.23

a raster image of erosion potential in the Rich Hollow basin (FIG 5). The areas that showed the highest susceptibility to erosion were the areas that were exposed soil. Fortunately, a large portion of this area had a land cover class other than exposed soil, greatly reducing its erosion potential. The average rate of erosion over this entire watershed was 7.48 t/ha/yr, which is considered high. The average rate of erosion for just the valley filled areas is much less with a rate of 5.22 t/ac/yr, a moderate erosion rate. A t-test was used to compare the erosion rates of the entire watershed and just the valley filled areas, and the resulting P value of 0.657 indicated that there was not a statistically significant difference between erosion potential of these areas. There were pockets of high erosion in certain areas that had a large influence on the average erosion rate. These pockets were especially prominent at the headwaters of the hollow, where the primary land class was exposed soil. Within these pockets, there were some pixels that had extremely high values, in some cases 1000 t/ac/yr and above. These are the sites that are most likely to be contributing heavy metal contaminants to Rich Hollow.

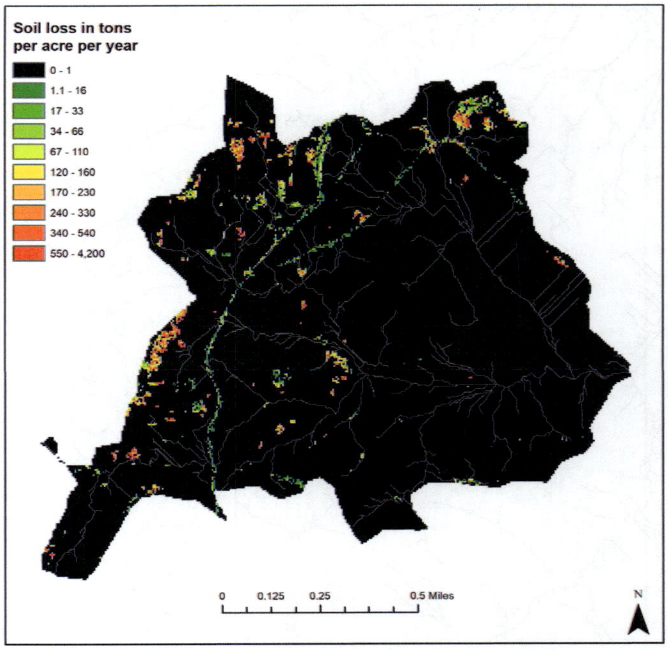

FIGURE 5. Soil loss in the Rich Hollow basin represented in t/ac/yr as calculated using the RUSLE method. Low values (black) represent areas of low erosion potential, while areas that are above 10 t/ha/yr (green) represent areas of high erosion potential. The highest values (orange and red) represent areas of extreme erosion potential.

DISCUSSION—This analysis was able to identify areas that are likely contributing to heavy metal contamination of the stream. This methodology is not a replacement for on-site measurements and analysis, but rather a tool for environmental restorationists to improve their efficiency and effectiveness. One of the most notable findings of this research was that the majority of the erosion at this site is not occurring on the valley fills. The biggest contributions of erosion are from small pockets of bare earth located at the headwaters of the stream. Even though these pockets are some of the furthest regions from where Rich Hollow drains into Big Horse Creek, they have the highest erosion potential.

These results could be supported by future research on site. Further sampling could be conducted to find the concentrations of contaminants on a finer hydrologic scale. By incorporating contaminant concentrations into a hydrologic model, the amounts of contaminant contribution from each erosion prone area can be calculated. This could be combined with the results of this study to more accurately locate point sources where the contaminants are originating. Further hydrologic modeling with network analysis could be used to assess the cumulative impact of multiple contaminated streams.

One potential issue encountered during the analysis was that the data were not all obtained at the same time. Because this mine site was changing so frequently from its inception in 1974 to 2015 when it was shut down, data obtained only a few years apart may have significant discrepancies. The area that would be most prone to this issue would be the western portion of the mine which experienced the most topographic change towards the end of the mine's lifespan. This may not have been a very significant issue within the Rich Hollow watershed because it was mined in the late 1980s.

In addition, this study did not take into account wind erosion because it has less of a direct impact on sedimentation delivery than overland flow. It would be valuable to account for this in future research because wind erosion contributes to the spread of dust. This has a direct impact on the communities surrounding the site, as well as any biological or human communities downwind.

CONCLUSION—By using the RUSLE erosion prediction method alongside hydrological analysis, scientists can estimate regions that are disproportionately adding heavy metal contaminants to the stream system in the Hobet Coal Mine in southern West Virginia. The primary factor that contributes to erosion at the Hobet mine complex is bare earth which is either exposed because of mining or from roads. Knowing where areas of high erosion risk are located can help ecological restorationists identify problem areas of the mining site that can be targeted for further study or remediation. This can also help them determine the appropriate restoration techniques for preventing polluted water from entering the stream system. Understanding the connections between different biotic and abiotic factors with GIS models can help humans minimize the long-term impacts of extremely destructive practices such as mountaintop removal coal mining.

REFERENCES

Arnold, M. C., Lindberg, T. T., Liu, Y. T., Porter, K. A., Hsu-Kim, H., Hinton, D. E., Di Giulio, R. T. 2014. Bioaccumulation and speciation of selenium in fish and insects collected from a mountaintop removal coal mining-impacted stream in West Virginia. *Ecotoxicology* 23: 929–938.

Bernhardt, E. S., Palmer, M. A. 2011. The environmental costs of mountaintop mining valley fill operations for aquatic ecosystems of the Central Appalachians. *Annals of the New York Academy of Sciences* 1223: 39–57.

Demirci, A., Karaburun, A. 2011. Estimation of soil erosion using RUSLE in a GIS framework: A case study in the Buyukcekmece Lake watershed, northwest Turkey. *Environmental Earth Sciences* 66: 903–913.

EPA. 2018. Rainfall Erosivity Factor Calculator for Small Construction Sites. Retrieved from https://www.epa.gov/npdes/rainfall-erosivity-factor-calculator-small-construction-sites#getTool.

Fox, J. 1999. Mountaintop removal in West Virginia: An environmental sacrifice zone. *Organization & Environment* 12: 163–183.

Griffith, M. B., Norton, S. B., Alexander, L. C., Pollard, A. I., & Leduc, S. D. (2012). The effects of mountaintop mines and valley fills on the physicochemical quality of stream ecosystems in the Central Appalachians: A review. *Science of The Total Environment* 417–418: 1–12.

Hendryx, M., Holland, B. 2016. Unintended consequences of the Clean Air Act: Mortality rates in Appalachian coal mining communities. *Environmental Science & Policy* 63, 1–6.

Hudson-Edwards, K. A. 2003. Sources, mineralogy, chemistry and fate of heavy metal-bearing particles in mining-affected river systems. *Mineralogical Magazine* 67: 205–217.

Kim, Y. 2014. Soil erosion assessment using GIS and Revised Universal Soil Loss Equation (RUSLE)." University of Texas at Austin, 2014.

Kouli, M., Soupios, P., Vallianatos, F. 2008. Soil erosion prediction using the Revised Universal Soil Loss Equation (RUSLE) in a GIS framework, Chania, Northwestern Crete, Greece. *Environmental Geology* 57: 483–497.

Liu, G., Xu, W. N., Zhang, Q., Xia, Z. Y. 2012. Interrill and rill erosion on hillslope. *Applied Mechanics & Materials* 170–173: 1344–1347.

Lechner, A. M., McIntyre, N., Witt, K., Raymond, C. M., Arnold, S., Scott, M., Rifkin, W. 2017. Challenges of integrated modelling in mining regions to address social, environmental and economic impacts. *Environmental Modelling & Software* 93: 268–281.

USDA. 2018. Web Soil Survey. Retrieved from https://websoilsurvey.nrcs.usda.gov/app/WebSoilSurvey.aspx.

West Virginia Department of Environmental Protection. 2018. Data Download. Retrieved from http://tagis.dep.wv.gov/home/Downloads.

Yenilmez, F., Kuter, N., Emil, M. K., Aksoy, A. 2011. Evaluation of pollution levels at an abandoned coal mine site in Turkey with the aid of GIS." *International Journal of Coal Geology* 86: 12–19.

Tathāgatagarbha and *Ātman*: Self Where There is No-Self

Aaron Alexander Laughlin[1*]

KEYWORDS— *Buddhism, Hinduism, self, philosophy, religion,* ātman, tathāgatagarbha, *Buddha-nature, Buddha-mind, emptiness, no-self*

*Corresponding Author: aal343@humboldt.edu
[1]Undergraduate
Department of Religious Studies
Humboldt State University, Arcata, CA 95521

Laughlin, A. A. 2019. *Tathāgatagarbha* and *ātman*: self where there is no-self. *ideaFest Journal: Interdisciplinary Journal of Creative Works & Research from Humboldt State University* 3: 66–70.

Humans have long grappled with the question of the nature of our Self, defined here as the ultimate reality inherent to our individual being. Religious traditions can be a great place to look when attempting to understand this aspect of our humanity. Broadly speaking, when contemplating ideas of Self in Buddhism and Hinduism, the relationship between the Buddhist notion of Buddha-nature (*tathāgatagarbha*) and the Hindu notion of Self (*ātman*), is an intriguing one: How can we understand them to be similar or different? How do the Buddhist concepts of emptiness (*śūnyatā*) and mind-only (*cittamātra*) relate to the concepts of *tathāgatagarbha* and *ātman*? Is emptiness contrary to these ideas? Are *tathāgatagarbha* and the Hindu teaching that *ātman* is equal to brahman (ultimate reality), both expressions of a non-dualistic state of mind? Although it is commonly taught that Hinduism and Buddhism differ in their understanding of Self, one thing that becomes apparent is that these are not simple questions, perhaps mainly because their answers are contextual. There are many answers that come from many different types of Hindus and Buddhists in various places. For this paper, I will be looking at commentary on the Buddhist text the *Mahāyāna Mahāparinirvāṇa sūtra* and its use of the concept of a permanent Self and how this relates to emptiness (*śūnyatā*) and skillful means (*upāya*). This paper seeks to support my claim that, through skillful means, *ātman* and *anātman* (no-Self) are both saying something quite similar—despite the apparent paradoxical nature of this statement—and will look at Buddha-nature in the *Mahāparinirvāṇa sūtra* as a way to understand and help articulate this thought.

I would like to make clear at the onset that I do not wish to claim an authoritative understanding of these concepts for Buddhists and Hindus (or anyone for that matter); these observations come from my own context and are not meant to speak for the traditions which I am observing. I am not a Buddhist nor a Hindu but I am intrigued by these concepts and wish to better understand myself and the worlds from which they come. This brings me to a question that has arisen for me during my research: why is Buddha-nature and *ātman* important to consider? I believe that the implications of these ideas have very tangible consequences in the world and can change how we think of concepts such as compassion. They can shape the way we contextualize ourselves in the world.

We must become comfortable with paradox when considering these ideas. Ultimately, we are using words as a means to an end (that perhaps is not to be thought of as an end exactly) that does not necessarily reflect the means. In other words, describing states of mind and ontology through words in these traditions results in contradiction, because words are limiting boxes that cannot completely contain or reflect reality. Yet, is it not these contradictions that bring us into a deeper interaction with these Buddhist teachings? Externalizing our search for understanding with words such as "deeper" help to illustrate this tension when looking at Buddha-nature; is there a "deeper" or "True" permanent Self, or is there no essential substantial Self in existence? Perhaps there is both. When speaking of the Buddha as seen in the *Mahāparinirvāṇa sūtra*, Paul Williams says, "He has taught Self where there is really no-Self, and no-Self where there is really

Self. This is not false but skillful means. Here the Buddha-nature is really no-Self, but it is said to be Self in a manner of speaking" (99).

The difficulty in using language to describe the tension between Self and no-Self can be seen in the *Nirvāna Sūtra* (as the *Mahāparinirvāṇa sūtra* has been referred to) when it asserts that the "core nature of each individual is that of a buddha, but mental afflictions (kléa) prevent most from realizing it" (Blum, 605). The semantic nature of using words such as Self and no-Self could be seen to be "mental afflictions" that prevent us from seeing that there can be a "core nature" and no core nature at the same time. Each of these teachings, Buddha-nature or *ātman* and emptiness or *anātman*, can steer people towards different attachments. When one does not see Buddha-nature it could be that attachment to the idea of an *ātman* could ultimately deter them from realizing their inherent Buddha-nature. Yet, this goes both ways; sometimes the best way to teach could be to teach of Self, in order to lead one in the most beneficial direction. Scholar Takasaki Jikidö reminds us that teachings are a means to liberation, "I would emphasize that the follower of the *Tathāgatagarbha* theory would be content with the evaluation of this teaching as "conventional," because any teaching of the Buddha is, after all, a convention or means for the sake of deliverance or religious awakening" (82). A conventional teaching may not reveal the precise ontological distinctions unique to each group of Buddhists that allows a Self to exist in a worldview that is empty of Self, but it shows that some Buddhists believe in the benefit to teaching Self, despite their apparent foundational idea of *anātman*.

One does not have to look too far into the traditions of Hinduism to see teachings that reflect the *Mahāyana* Buddhist idea of the existence of a core nature and no core nature at the same time. When the *Mahāparinirvāṇa sūtra* speaks of our core Buddha-nature and the Hindu texts the *Upaniṣads* speak of *ātman*, there are parallels between the two. In the *Mahāparinirvāṇa sūtra*, there is a story about a king who seeks the sound of a lute. The king futilely attempts to find the sound in a lute by breaking it apart until a minister explains to him that this is not the way to get to the sound. In *The Doctrine of the Buddha-Nature* in the *Mahāyāna Mahāparinirvāṇa-Sūtra* by Ming-Wood Liu, the author explains the themes in this story as they relate to Buddha-nature, "The central theme of the story is summed up in the concluding declaration that the Buddha-nature "abides nowhere," i.e., is not immanent in some form in sentient beings, just as sound is not immanent in any part of the lute. In the same manner as sound is produced when all necessary conditions are satisfied, the Buddha-nature will reveal itself to sentient beings when they practice in earnest the way to enlightenment prescribed by the *tathagata*" (Liu, 82). This story resonates strongly with one from the *Chandogya Upaniṇad*. In this story, Śvetaketu is being taught by his father about the nature of *ātman* and *brahman*. The father asks Śvetaketu to divide a fig, and then a seed from the fig, and when Śvetaketu says that he sees nothing by dividing the seed, the father says, as quoted in *A Survey of Hinduism*, "'My dear, that subtle essence which you do not perceive, that is the source of this mighty Nyagrodha tree. That which is so tiny is the *ātman* of all. This is the true, the self, that you are, Śvetaketu'" (Klostermaier, 169). Both stories could be seen to speak of a self that does not exist inside of things, and cannot be found by breaking things down. Yet, it is the breaking down of the lute and the fig that aid in the acknowledgement of the ideas of *ātman* and Buddha-nature because there is something else there that has yet to reveal itself. The essence that cannot be seen is that which is doing the seeing; the unseen seer or *ātman* and buddha womb or *tathāgatagarbha*, could be understood through these stories to be consciousness itself.

Both Hindus and Buddhists have described our core nature, or Self, as an unadulterated state of consciousness. The *Mahāparinirvāṇa sūtra* describes Buddha-nature as the pure, inherent element within everyone that is the potential for one to attain buddhahood, which could be interpreted as to imply that it is consciousness. Later *sūtras*, such as the *Śrīmālā sūtra* describe *tathāgatagarbha* as "intrinsically pure radiant consciousness (pp. 106–7)" (Williams, 102). *Ātman* is described in the same way. According to *A Survey of Hinduism*, "*Ātman* is pure consciousness" (Klostermaier, 357). Consciousness is a universal phenomenon through which a myriad of manifestations has expressed themselves throughout time; it is quite possible, from my perspective, that many of the arguments and distinctions between Buddhist conceptions of no-Self and Hindu conceptions of Self have been the result of getting caught up in words because semantics can get sticky; words are messy, impermanent, and hold the power to nudge towards enlightenment but also confuse and separate people. Through the eyes of the *Mahāparinirvāṇa sūtra*, the Buddha did not care enough about the doctrine of no-Self to defend it needlessly; when accused of nihilism by Hindu *Brahmins* he responded by affirming his teachings as describing a Self. This portrays a value for flexibility over rigidity as well as a sense of skillful means and a compassionate, enlightened perspective.

It has been argued in Buddhist *sūtras*, such as the *Śrīmālā sūtra*, that it takes an enlightened perspective to accurately speak of Self in a reality of no-Self. The *Śrīmālā sūtra* states that, "it is difficult to understand the meaning of the intrinsically pure consciousness in a condition of defilement" (Williams, 102). This seems to be at the crux of the apparent issue between *ātman*, *anātman*, and Buddha-nature; Buddha-nature is not in the descriptions of it, but in the pervasive sound that coalesces from an understanding free of descriptions. This train of thought—which is really a kind of no-thought—can be seen to go back to the *Upaniṣads*, as quoted in *A Survey of Hinduism*, "'This Self cannot be attained by instruction, nor by much thought nor by listening to many scripture readings: the Self is only attained by one who is chosen: to such a one the *ātman* reveals itself'" (Klostermaier, 172). Much like in the story of the king and the lute, Buddha-nature reveals itself when one is, in a way, chosen or ready to see that which is waiting to be revealed. Hindus and Buddhists have both understood the Self in an empty way; the Self is revealed when one finds their way out of the endless vortex of descriptions that try to describe itself. The Self exists within a context of no-Self.

By looking at the story of Yājñavalkya and Gārgī Vacaknavī in the *Bṛhadāraṇyaka Upaniṣad* one can see how Self can exist within a context of no-Self. Through this story we can see how the idea of *ātman* and brahman could be seen to exist in emptiness, which could mean that *ātman* does not have to be at odds with *anātman*. At one point, Gārgī questions Yājñavalkya about what the world is woven on warp and woof, in other words, what are the frame and threads that make up the fabric of our reality. Eventually, after going from water, to wind, to creation and so on, her questioning arrives at *brahman*, which she also questions, asking for its source as well. Yājñavalkya stops Gārgī's questioning here saying, "'Gārgī do not question too much lest your head fall off. You are questioning too much about the divine being, about which we are not to question too much" (Klostermaier, 167). To me, this story implies that one could keep questioning and breaking things down to further realize the extent of emptiness, but Yājñavalkya recognizes this as unhelpful. Using skillful means, he stops her questioning at *brahman*; this not only affirms that *brahman* is the level of reality that Gārgī (at this juncture in her understanding) should be concerned with, but also affirms the ambiguous nature of this teaching. The teacher uses skillful means in this story to end the questioning mind of someone perhaps not ready to contemplate the extent of emptiness. Much like the affirmation of mind in *Yogācāra* Buddhism in the face of emptiness, *ātman* affirms a quality much like Buddha-nature in the face of the knowledge that there is always more to dissolve under analysis. The question becomes what is the use, or skillful means of dissolving everything? What is the need or benefit of going beyond *ātman*? If the danger in *ātman* is attachment to the idea of a Self, which is ultimately an illusion or egoistic self, how different is it to be attached to the notion of Buddha-nature or *anātman*? One answer is as follows, "The theories of *ātman* and *anātman* are both "skilful ways" (*upāyah*) to save ordinary men from errors. Neither *ātman* nor *anātman* are the truth" (Ishigami-Iagolnitzer, Mitchiko, 5). My interpretation of this quote tells me that we should appreciate the deep ambiguity of these notions and their ability to adapt to the needs of the one investigating them.

The Hindu notion of brahman portrays this deep ambiguity, which is characteristic of the conventional teachings we have explored within Buddhism. Since brahman is understood to be beyond even creation itself, it can be said that one could conceive of it as a no-thing that is empty of our conception of it. The name brahman could be understood as a conventional means of explaining something beyond words. As quoted in *A Survey of Hinduism*, the *Upaniṣads* say, "Where words do not reach and the mind cannot grasp, there is the brahman full of bliss" (Klostermaier, 168). Scholars and *Mahāyāna* Buddhists have recognized the similarity here to Buddha-nature thought. "It seems to be a return to the *ātman* (or *Brahman*), but this Great Self, for *Mahāyāna* Buddhists, is only a conventional name, given to reality void of substance, which is Vacuity and *Nirvāṇa*" (Ishigami-Iagolnitzer, Mitchiko, 5). Both traditions are pointing towards a unifying substance with no substance; a consciousness free of deluded consciousness described by words that are empty of ultimate meaning or truth, in order to indicate a suchness in nothingness.

I find it important to note at this point in the paper the fact that religious traditions change, and our perception of them should keep this in mind. Ideas develop as time passes and outside influences affect the traditions and ideas within them. Buddhism has responded to itself and attempted to reconcile seemingly contradictory ideas, such as the emptiness described by *Madhyamaka* teachings with the mind-only teachings of *Yogācāra* and *anātman* with Buddha-nature. In the lengthy paper entitled *A Study of Yogācāric Influence on Tathāgatagarbha Doctrine as Found in Lahkavatarasūtra*, by Mei Hsiao, this change is examined in detail:

> "Finally, having thoroughly examined *tathagata-nairatmya-garbha* in the *Lankdvatdrasūtra*, it was found that the doctrines of *tathāgatagarbha* and *pudgala-nairatmya* were aligned with each other but only under a certain condition—that is, only when the *ātman* proposed by other religions was denied. However, from the viewpoint of the metaphysical aspect of the *Lankdvatdrasūtra*, the *tathāgatagarbha* can be considered to be the genuine *Ātman*, but one which is very different from the absolute *an-ātman* declared in Primitive Buddhism. Actually, there is a noticeable inconsistency between the views of Primitive Buddhism and the *tathāgatagarbha* tradition" (Hsiao, 69).

While my paper is generally in disagreement with the notion that the *tathāgatagarbha* tradition is inconsistent with the *anātman* of so-called "Primitive Buddhism," I find, as discussed throughout this paper, that through the example of the *Mahāparinirvāṇa sūtra* one can see how these ideas are saying something similar as a form of paradox. It is important to make note of the fact that there are many ways people have viewed this problem. The focus of my paper is not to detail the historical setting that influenced the development of these ideas; there are many social and political factors that have greatly affected the development of Buddhism and Buddha-nature and they are all important to consider. But, I believe there is a way to see a connection between Buddha-nature, *ātman*, and *anātman*, and a text such as the *Nirvāṇa Sūtra* is a great place to observe this train of thought. Hsiao sees this as well as a clearly popular trend among later *Mahāyāna* Buddhists, and states that "In the ideological trend of later *Mahāyāna* Buddhism, the doctrine of 'real and eternal mind-only' became all influential and dominant" (Hsiao, 81).

When reading through *A Study of Yogācāric Influence on Tathāgatagarbha Doctrine as Found in Lahkavatarasūtra*, it became clear that there has been much justification of Buddha-nature thought by Buddhists, some that accept *ātman* within their tradition and some that do not. The philosophy of *ātman* was considered by some to be heretical, "The compound "*tathagata-nairatmya-garbha*," according to the context from which it is extracted, aims to indicate that the teaching of *tathāgatagarbha* is entirely different from the theory of *ātman* held by the heretical philosophers" (Hsiao, 42). There are many reasons as to why this may have been the case and among them could be that there is a lot at stake (socially, philosophically, politically, and so on) in maintaining the distinction between Buddhism and Hinduism, which can hinge upon the distinction between *anātman* and *ātman*. Part of my claim, and support for a text such as the *Nirvāṇa Sūtra*, is that this goes against an understanding of skillful means. As a result of acting within the world, these Buddhists may be attempting to hold onto their traditions and identities as forms of attachment (for good, valid reasons maybe), but this, perhaps, leads one away from the ideals of Buddha-nature. My claim here is not some sort of ultimate truth however, and I am in no way saying I am righter than Buddhists (and non-Buddhists) who follow this thought pattern or that this thought is not valid or correct from some perspectives. The world is incredibly complex and this is simply an opinion based on my own relative understanding.

Although there may not be one true way to conceive of the Self through the lens of Buddhism and Hinduism, I believe there is great benefit in placing these ideas within a context of compassion. In the dissertation by Kiseong Shin called *The Concept of Self and its Implication for Salvation in Hinduism, Buddhism, and Christianity*, when discussing the implication of the ideas of self in *Advaita Vedānta* the author states, "Harmony and balance of the universe is intrinsic because all things are united together in one true self" (Shin, 184). This is certainly one way we can take the teachings of *ātman* and *brahman*. It is important to note, however, that historically speaking this is not the only way these teachings have been taken. If *ātman* is eternally pure and cannot be defiled then one could conclude that death, whether inflicted towards oneself or another may not, in a sense, be significant. That being said, this is not the only necessary conclusion either. My point here is that the realization of Self in the form of *ātman* or Buddha-nature is not inherently compassionate. Many times in Hindu mythology yogis receive great destructive power from deep realization and as seen in the recent conflict in Myanmar, Buddhists are certainly capable of violence as well. Yet, conceiving of Self in the ways understood through this paper can be an incredibly compassionate ideal, as Buddhists have shown through their teachings. Shin sees similarities between Buddhist and Hindu thought and concludes in one section of the paper on Buddhism that, "Compassion is essential in the realization of co-origination of everything because everything is interconnected with everything else. Bowers argues that everything in the world is co-originated, and self and other are non-differentiated, then 'loving the other means loving the self'" (Shin, 187). When considering this within my own context, I share the same sensibilities and find the careful exploration of these ideas to be a valid and thorough way of fostering a compassionate state of mind that can directly influence one's behavior in the world.

Throughout the process of writing and researching for this paper, I have found the theory of Buddha-nature as Self to be a beautiful idea. Maybe this could be explained because I have an attachment to the idea of a Self and this is ultimately a hindrance on my own potential path to a less deluded realization of reality. Perhaps Self for me could be a beneficial convention to deepen my own understanding. I stand open to the possibility of abandoning my own affinity for a pure, permanent form of (or experience of) consciousness that is the nature of one's Self and true reality (that cannot be truly reached through language). However at this point, for my personal understanding, I prefer to find a similar beauty in the theories of Buddha-nature and *ātman* without saying they are or must be considered identical. I find this beauty to be an important and valuable thought; one that can help the world by helping to invigorate a sense of beauty, acceptance, and compassion.

ACKNOWLEDGMENTS—The author would like to thank Prof. Stephen Jenkins, as well as professors Sara Hart, Katherine Wilson, Jeff Sellars, William Berbrechtmeier, and Karen Harris.

REFERENCES

Blum, Mark L. "Nirvāṇa Sūtra." Encyclopedia of Buddhism. Ed. Robert E. Buswell, Jr. Vol. 2. New York: Macmillan Reference USA, 2003. 605-606. Gale Virtual Reference Library. Web. 27 Feb. 2017.

Grosnick, William H. "Tathāgatagarbha." Encyclopedia of Buddhism. Ed. Robert E. Buswell, Jr. Vol. 2. New York: Macmillan Reference USA, 2003. 826-828. Gale Virtual Reference Library. Web. 25 Feb. 2017.

Hsiao, Mei. "A Study of Yogācāric Influence on Tathāgatagarbha Doctrine as found in "Laṅkāvatārasūtra"." University of Calgary (Canada), 2008. Ann Arbor: ProQuest. Web. 25 Feb. 2017.

Ishigami-Iagolnitzer, Mitchiko. "The Self and the Person as Treated in some Buddhist Texts [1]." Asian Philosophy, vol. 7, no. 1, Mar. 1997, p. 37–45.

King, Sallie B. "Buddha Nature and the Concept of Person." Philosophy East and West, vol. 39, no. 2, 1989, p. 151–170.

Klostermaier, Klaus. A Survey of Hinduism. 3rd ed. Albany, NY: State U of New York Press, 2007. Print.

Liu, Ming-Wood. "The Doctrine of the Buddha-Nature in the Mahāyāna 'Mahāparinirvāṇa Sūtra'." The Journal of the International Association of Buddhist Studies, 5.2 (1982): p. 63–94.

Shin, Kiseong. "The Concept of Self and its Implication for Salvation in Hinduism, Buddhism, and Christianity." Drew University, 2013. Ann Arbor: ProQuest. Web. 25 Feb. 2017.

Takasaki, Jikidō. "The Tathāgatagarbha Theory Reconsidered: Reflections on some Recent Issues in Japanese Buddhist Studies." Japanese Journal of Religious Studies, vol. 27, no. 1–2, 2000, p. 73–83.

Williams, Paul. Mahāyāna Buddhism: the doctrinal foundations. London: Routledge, 2003. Print.

Rhetorical Genre Theory and Whiteness

Greg W. Childs[1*]

KEYWORDS—*rhetorical genre theory, whiteness, composition, first year writing, white supremacy, higher education, James Baldwin*

INTRODUCTION—Whiteness is a socially constructed system of norms that grant those classified as white—typically individuals with light skin—a great deal of unearned opportunity and privilege. The definition of what makes someone white is negative, that is to say, it is based on what they are not rather than what they are. To be white is to not be a person of color; the grouping of people who are white do not have anything in common ethnically. In the United States whiteness is used as a way to maintain white dominance over a society that claims to be a meritocracy. Those in the white group are privileged, with their ways of knowing and being considered norms. All others must conform or suffer the constitutions of being rejected from the dominant group.

It is also important to understand that whiteness intersects with class status. Whiteness and the American middle class are linked by common values and exceptions. If one can navigate the ways of knowing and being of whiteness, they can likely navigate our middle class. It is not a coincidence that for a child of white parents the path through the educational system and into the middle class is relatively unobstructed. Thus they not only end up in a position of relative socioeconomic power, but also reinforce the myth of the meritocracy in their own minds.

Whiteness is the foundation on which academia is built. It shapes the institutions and methods of knowledge-making that form what we call education. If we hope to make radical, meaningful change to our systems of learning and knowledge, a critical step will be decentering whiteness. In this paper I start by discussing and marking my own whiteness and describing the ways in which it has influenced my experiences in higher education. I do so to clarify that this is a white, middle class, and male piece of writing—identities that all too often go unstated and unaddressed by the author. This lack of awareness and discussion of an author's whiteness contributes to fortifying the hegemony of whiteness.

I also define whiteness as I am conceptualizing it in this paper. To help define what it means to be white and how it intersects with learning, I look to James Baldwin's thoughts on whiteness and integrate his writing throughout the paper. It is also important to address how writing instruction serves as a gatekeeper of the academy, privileging the thoughts and practices of those who have access to white, middle-class discourse, and how white teachers fall into the trap of reinforcing white hegemony.

Finally, I address rhetorical genre theory as an example of how, like many of our tools and methods of knowledge-making, it arose from whiteness and should not be considered ideologically neutral or universal. In closing, I further discuss what Baldwin termed the "price of a ticket" to access whiteness and outline some of the work that the individuals most privileged by the hegemony of whiteness must do to aid in its decentering.

MY WHITENESS—I have had what many would consider a successful but unexceptional college experience. Checking the boxes, I worked my way from remedial classes at community college to graduate- level courses during my last year as an undergraduate. One of the things I want to examine in this paper is why my experience, as a white, middle-class male, is considered so typical that no one would think to look more closely at it. What allowed me to go from remedial classes to doing graduate coursework as an undergraduate? As much as I might be encouraged to say that it was only my hard work and dedication, and to think nothing more about it, I don't believe it was. I know that I was able to succeed

*Corresponding Author: gwc15@humboldt.edu
[1]Undergraduate Student
Department of English
Humboldt State University, Arcata, CA 95521

Childs, G. W. 2019. Rhetorical genre theory and whiteness. *ideaFest Journal: Interdisciplinary Journal of Creative Works & Research from Humboldt State University* 3: 71–76.

because I am operating in institutions that were built from the ground up for people who look and think like me—people whose families hold the same values and have the same histories as mine. Access to whiteness has influenced my success in college. I was supported by a system of norms that has allowed me to excel because I was born into it, raised in it—it is the air I live and breathe. And just like the air around me, I can't see it, can't feel it; no one is going to point it out to me, and say "You've done well, but be sure to thank the invisible force guiding you along the way." If someone said to me that it was not my brain, but my whiteness that got me to where I am, it would be culturally normal for me to take offense. Those of us who are white have been trained to defend its invisibility. We do this by focusing on the white experience as a normal and neutral baseline, and disregarding or disbelieving the experiences of those who are not white. But I have come to understand the importance of naming whiteness for what it is and to always question my idea of normal in any given setting.

It would be easy for me to think that the journey I have had is typical or to think it was not anything outside of my control that facilitated my success and ease of adjustment in higher education. But I have learned better than that. I now know that I am the type of person these institutions were built for, built on an academic tradition going back to the abbeys of Europe. While I may have had my own private doubts about my place at Humboldt State University, it is unlikely anyone else doubted me.

My identity as a white, middle-class male is an important influence on my interpretation of whiteness in academia. Without bringing my identity to light, it would fade into the background as it is the dominant, "neutral" discourse. The white man is the default, the assumed identity. Those with identities like mine are granted ideas without scrutiny, and they enter the scholarly conversation without an attached qualifier, such as an LGBTQ author or writer of color. It may seem odd for a white person to argue for the decentering of whiteness, however, I believe that whiteness in academia cannot be decentered if it is not addressed. Just like any other part of a person's identity, having access to the power structure of whiteness shapes how the world is seen.

People who grow up white rarely have their view of the world challenged. It often seems to them that everyone sees and experiences the same world, regardless of whether they are white. They also tend to be unaware of the influence that their whiteness has on their view of their environment. White people can easily believe that they see the world objectively, that there are no parts of their identity that are interfering with their assessment of reality. Robin DiAngelo (2016) explains it well through her experience in her book, What Does It Mean to Be White? Developing White Racial Literacy: "I did not see the world objectively as I had been raised to believe, nor did I share the same reality with everyone around me. I was not looking out through a pair of objective eyes, I was looking out through a pair of white eyes" (p. 2). This belief is reinforced by a white dominant system that allows white people to remain immersed in whiteness in their daily lives, including in academia. Throughout my years at Humboldt State University, for example, all of the classes I have been required to take have been taught by white faculty. In the fall of 2016 Humboldt State University had 241 tenured/tenure-track faculty, 77% of whom were white. The English department I study in reflects these numbers. This is not uncommon in higher education and only helps make reality seem more normatively white.

WHITENESS AS POWER—I conceptualize whiteness not as a race but as a power structure. Within this structure those with white skin are given more privilege, opportunity, and resources. These advantages are not acknowledged. The success of those with access to whiteness is used to measure what is normal within the structure, while those outside of the white group are deficient. Whiteness, and the understandings it holds as norms, has become a powerful hegemonic force in the world around us. These understandings are reinforced at every level of education. In schools, and in higher education especially, whiteness acts as a gatekeeper of success. Those who cannot make it in school are written off by society as hopeless, and the white, middle-class dialect used in schools is a tool to dictate who reaps the rewards of education. The fact that high school dropouts make so much less than college graduates is framed not as a social injustice, but as punishment for those who could not succeed in the educational system and a reward for those who could. This power structure does the work of both making sure that those who succeed already have access to whiteness and of blocking the progress of those that do not, thus perpetuating power imbalances between those with access to education and those without.

James Baldwin was a novelist, essayist, and social critic who wrote at length and with piercing insight about the American experience. Baldwin articulated the white experience as lacking both culture and ethnicity, an experience that is nothing more than an invisible power structure that privileges those deemed worthy of access. Baldwin (2011) wrote of whiteness: "The world is not white; it never was white, cannot be white. White is a metaphor for power, and that is simply a way of describing Chase Manhattan bank" (p. 158). Baldwin's use of whiteness as a metaphor for power has been influential in my own conceptualization of whiteness. Stating emphatically that the world is not white, cannot be white, speaks to the fact that no matter how deeply entrenched whiteness might be, it is nothing more than a way of seeing the world that is invested in maintaining the power held by the likes of Chase Manhattan bank. If someone is white, they are not part of a group defined by anything but power. The concept of whiteness is socially constructed to privilege and oppress, and it has become an incredibly strong hegemonic force in our world. Those with access to whiteness are trained to see their privilege as innate superiority and personal achievement, while those without access are treated as deficient or personally responsible for their oppression. Conceptualizing whiteness as power helps it become visible. It traces and decenters the hegemony of whiteness and makes visible and dismantles the ways in which it supports white supremacist power structures within our institutions.

We do not have to look far to see the influence of whiteness in our educational system. Differences in outcomes between white and non-white students are referred to as "achievement gaps"—the white students' level of achievement is the norm while all others' underperform. The underlying influences of whiteness must be addressed and made clear that the norms it holds are not universal or superior. As James Baldwin wrote: "Not everything that is faced can be changed; but nothing can be changed until it is faced" (2011, p. 42). The ways in which the hegemony of whiteness is taken for granted in education must be faced if it is to be changed.

Baldwin (2011) also articulated the price of whiteness and how America became white: "The price of a ticket was to cease being Irish, cease being Greek, cease being Russian, cease being whatever you had been before, and to become 'white.' And that is why this country says it's a white country and really believes it is" (p. 156). The price of a ticket, as Baldwin puts it, is to give up who you are. This is fine for those of us born into whiteness; we never had cultural attachment to begin with. But the cost is much higher for those who must cease being who they are culturally and survive in a white supremacist system by assimilating to a view of the world that does not make room for them.

WHITENESS AS OBJECTIVE AND NEUTRAL—Whiteness and langue are closely related to each other. "Standard English"—the English canonized in many style guides and manuals—is the langue of whiteness. It is the primary dialect of the white, middle-class people in the United States. Even though The Conference on College Composition and Communication has asserted Students' Right to Their Own Language (1974), it is still enforced as proper and the norm. Those who speak English get a head start on their education, as they are not being corrected by their teachers for the way they speak. An example of how teachers may reinforce whiteness in education is offered by Joan Wynne (2008), who writes about her experience giving questionnaires to gauge the attitudes of almost uniformly white teachers-in-training and working teachers about language use in the classroom. One of the respondents claimed that "all children should speak 'Standard English' because, 'We are a part of the Human Race and Standard English is the common denominator,' ... a 'neutral and universal language'" (p. 211–2). This response is one of the more extreme, but it reflects the overall response to Wynne's questions about language. Particularly telling is the comment about Standard English being "neutral" and "universal," as both terms accentuate the teacher's unawareness of their whiteness. This person does not see their whiteness as something that is politically charged or as something that gives them a great deal of social power. To them it is normal and natural, the "common denominator" across all students. They do not see the privilege embedded in that statement, to never having to think about their language as anything but universal. These teachers are the foot soldiers of whiteness. They guard the gates of social power and accessibility, and without knowing it beat back language expression that is not white and middle-class.

Writing classes are some of the most powerful tools that academia has to teach students the logic and methods of whiteness as neutral and universal. Even position statements like the Students' Right to Their Own Language (1974) do little to address the way white, and supposedly neutral and universal, academic discourse is used as a tool of white supremacy. It certainly is important to articulate that students have a right to their own language and composition studies have done good work to move the field away from skill and drill grammar instruction. But just as removing a dress code will not remove the stigma attached to some clothing, allowing students to use their own language does not remove the stigma attached to non-white ways of speaking. As Asao Inoue (2014) states, we must do more than simply acknowledge that there are many ways of speaking (p. 71), we must also stop believing that white discourse holds the keys to "creativity, insight, critical thinking, explanation, and communication." Language differences must not be viewed as a deficiency to be overcome (p. 89). Inoue highlights one of the ways that language is an effective gatekeeper for whiteness. Even when non-white discourse is an option for students, it doesn't change the sigma that other forms of language are less effective, less objective, or even less moral than white, middle-class English.

In "Freshman Composition as a Middle-Class Enterprise," Lynn Bloom argues that first-year college English classes are a site where middle-class values are instilled in students as they enter higher education. Bloom's writing focuses on middle-class values, but I think that it can also be interpreted as a statement on instilling whiteness in first-year writing classes. These writing classes are understood to be one of the key places where the skills that are needed to succeed in college are taught. In an institutional system that is so deeply entrenched in whiteness, it is not hard to imagine that these first-year writing classrooms might become one of the places where whiteness is reproduced in the academy. The so-called objective and neutral logic and practice of academia are shown to students, and the students are told that to succeed they must learn to use these tools. Or as Bloom (1996) put it more bluntly: "Like swimmers passing through the chlorine footbath en route to plunging into the pool, students must first be disinfected in Freshman English" (p. 656). The danger of this process is that it is seen as ideologically neutral. Students are learning to be objective and think critically, but they are doing so on the terms of an educational system that is based in whiteness. Because the whiteness of the academy is unmarked, the tools that students are expected to use are working on students as much as for them.

RHETORICAL GENRE THEORY AND WHITENESS—Whiteness plays a central role in shaping the academy, especially in the first-year writing classroom. It would be worthwhile to look more closely at one of the theories that informs those writing classes and how they are both shaped by whiteness and help explain how whiteness shapes the academy. Rhetorical Genre Theory is a way to understand rhetorical situations. This theory holds that "genre constructs and responds to recurring [social] situation[s]" and that "Genre is truly … a maker of meaning" (Devitt, 1993, p. 580). Or as Bawarshi (2000) put it: "Genre is what it allows us to do, the potential that makes the actual possible, the 'con' and the 'text' at the same time" (p. 357). It helps to understand the rhetorical situation and how it functions within our society, but it also illustrates that genre is contingent upon the social situations that arise. Genres both meet the needs of the situation and shape the response to it. We write thank you notes because of a need to express gratitude, and the note we write is shaped by the rhetorical constraints put in place by the genre. Miller (1984) discusses how those situations are constructed: "Situations are social constructs that are the result, not of 'perception,' but of 'definition.' Because human action is based on and guided by meaning, not by material causes, at the center of action is a process of interpretation" (p. 156). If human action is directed by meaning, and genres are makers of meaning, then genre is a critical piece to understand when considering how writing is constructed, how it works in the world after its construction, and how these two phases are interrelated.

But how does whiteness work within this theory that can tell us so much about the social functions of writing? Miller (1984) addressed the fact that there are many ways recurring situations could be reacted to and lamented the lack of shared meaning among individuals: "What recurs for me does not for someone else; with a wealth of stimuli and a dearth of shared knowledge, we hardly know how to engage each other in discourse. We have many and confused intentions, but few effective orientation centers for joint action" (p. 158). Bawarshi (2000) frames Genre as a tool that has the power to bridge differences in the field of English studies: "I posit genre theory and analysis as a method of inquiry that might very well help us synthesize the multiple and often fractionalized strands of English Studies" (p. 336). The kind of universalizing being called for would be useful in the sense of increasing the potential for shared knowledge and community action. But it is important to ask, on whose terms are methods and knowledge being universalized? A lack of shared meaning should not be solved by developing a normalized understanding of the world based in whiteness. This is not to say that rhetorical genre theory is not useful or cannot lead to greater understanding. But tools used to gain understanding cannot be viewed as neutral, reflective of a universal, ideologically neutral form of knowledge. This problem is compounded by the fact that a white lens, the functional foundation of the academy, is conflated with a lens that is objective and neutral. Universalized knowledge based on a white lens would only serve to reinforce white supremacist hegemony. While knowledge-making or inquiry of any kind always employs some type of lens, it is important that these lenses are discrete and obvious tools. White scholarship and white understandings of the world must be seen as white, not as neutral and universal.

Another example of the universalizing of rhetorical genre theory is given by Bawarshi (2011), who posits what he calls the genre function: "which constitutes all discourses' and all writers' modes of existence, circulation, and functioning within a society" (p. 338). Genre may indeed contain all of these things, but the theories used to explain genre come from a small set of academically located discourses that are centered in whiteness. It is important to make the distinction between a tool of inquiry and an object of study. The former is the theories and methods used in academia to make sense of the world; the latter is the part of the world being studied. These tools are no more neutral nor objective than the world they seek to understand. The ways of knowledge-making we have at our disposal may seem to account for everything, but they can only complete this task by using the basic assumptions of the hegemony of whiteness.

An issue that arises with analytic tools rooted in whiteness is that while they may expand understanding for those with access to whiteness, they limit other ways of understanding. When whiteness is presented as neutral or universal, it cuts down the potential for a plurality of voices. Theories based in whiteness limit our tools of understanding. Miller (1984) states that, "The number of genres current in any society is indeterminate and depends upon the complexity and diversity of the society" (p. 163). However, the academy works against the complexity and diversity of the students who enter and aims to make them into scholars who will develop the complexity of white and middle-class ("academic") thought and theory. This means that the academy is a self-regulating system that limits processes of understanding to standards of scholars based within whiteness. Students must both be given access to the knowledge-making tools of the academy and encouraged to fully understand the ideological nature of those tools.

The paradox of using tools based in whiteness to decenter whiteness must be addressed head on. Intersectional feminist Audre Lorde (1984) wrote: "The master's tools will never dismantle the master's house. They may allow us temporarily to beat him at his own game, but they will never enable us to bring about genuine change." This could also be interpreted as the idea that the tools of whiteness will never decenter or dismantle whiteness. It is important to remember that rhetorical genre theory is both a tool of whiteness and a tool that can be used to help understand whiteness. But the paradox of using a tool of whiteness to work against whiteness should not be forgotten.

Rhetorical genre theory is also connected to whiteness in that it searching for typicality among texts. Rhetorical theorists are likely to be seeking conformity and examining generic expectations (Devitt, 2000, p. 705). The fact that whiteness plays such a large role in shaping the typical genre means that what rhetorical genre theory marks as typical will likely be white and middle class, and what it marks as deviations or errors will not be. This is not to say that identifying and understanding genre with the tools of rhetorical genre theory cannot teach us a great deal, but the positionality of the tools themselves must be acknowledged. It must be understood that the standards and expectations of academic theories are not ideologically neutral. When the theories being used are meant to explain how rhetorical situations are shaped and met, it is important that the forces working to shape those situations are being accounted for. This is where whiteness becomes an important factor. It is the power structure that shapes so much of what we do and how we think in the academy, and we must make ourselves aware of how it influences what we write, say, and think. This is especially true in a setting that gives teachers the power to set generic expectations, and the students' ability to meet those expectations has a great impact on their success or failure.

Charles Bazerman (2004) addresses the need for a greater understanding of how genres function for different populations directly: "There are serious methodological difficulties with relying totally on our 'native speaker intuitions' as anything more than a first approximation. Technically, relying on our intuitions already makes us assume many of the things we want to investigate." What Bazerman calls "our native speaker intuitions" could also be called our whiteness. He goes on to state that "We are already assuming that everybody understands these texts exactly as we understand them" (p. 377). The understanding that Bazerman references is rooted in the fact that the tools being used are built out of whiteness and supported by it. Just as genre reacts to and shapes rhetorical situations, so too does whiteness, as it works to shape assumptions in ways that both make it appear to be objective and reinforce white supremacist hegemony. It is not enough to make gestures acknowledging whiteness, to point out that we are making assumptions based on whiteness. Whiteness must be actively worked against by exposing and decentering it. The ways it shapes our reality must be understood and pointed out at every opportunity.

CONCLUSION—Rhetorical genre theory is just one example of how the tools we use to generate knowledge are embedded in whiteness. The basic assumptions of whiteness are what allow for the theories and methods of academia to be seen as universal. The question becomes, how can we decenter whiteness in our thinking and research? The first step is identifying whiteness, making it visible, working against all that has been done to make it as invisible as the air around us. Bazerman (2004) suggests responding to what he calls relying on "our native speaker intuitions" with the following:

> We have only a bootstrapping operation of increasing our knowledge and perspective through research such as examining more texts in a more regularized way; interviewing and observing more writers and readers, and ethnographically documenting how texts are used in organizations (p. 377).

There is value in expanding knowledge in the ways Bazerman is suggesting, but there is also a need to take a step back and be aware of the tools one is using. Leaning more heavily into methods embedded in whiteness, without acknowledging whiteness, will only lead to greater understanding through a white lens. The hegemony of whiteness will remain unchallenged and will continue to function as a white supremacist system of privilege and oppression.

Baldwin (2011) offered his own thoughts on how we might address whiteness in the academy and scholarship: "One hears from a long time ago that 'white is merely a state of mind.' I add to that, white is a moral choice. It's up to you to be as white as you want to be and pay the price of that ticket" (p. 157). There is a choice to be made when it comes to whiteness. It can be left unchallenged and unmarked, or it can be marked, decentered, and dismantled. Anyone who is situated within the "white state of mind," and who has access to the power structure that is whiteness must develop reflective habits and must be willing to stop paying the unseen moral price of membership. This price is not easy to become aware of, but it must be understood that the centering of whiteness at the exclusion of any other experiences will always lead back to a system in which the white, middle class, and male will be privileged. It is critical for those who are the most privileged by whiteness to do the introspective work required and learn how to stop blindly paying the moral price of whiteness.

ACKNOWLEDGMENTS—Thanks to Dr. Janelle Adsit for her guidance and feedback as I research and wrote this paper.

REFERENCES

Baldwin, J. (2011). The cross of redemption: uncollected writings. (R. Kenan, Ed.). New York: Vintage International.

Bawarshi, A. (2000). The genre function. College English, 62(3), 335. https://doi.org/10.2307/378935

Bazerman, C. (2004). Speech Acts, Genres, and Activity Systems: How Texts Organize Activity and People. In C. Bazerman & P. A. Prior (Eds.), What writing does and how it does it: an introduction to analyzing texts and textual practices. Mahwah, N.J: Lawrence Erlbaum Associates.

Bloom, L. Z. (1996). Freshman composition as a middle-class enterprise. College English, 58(6), 654. https://doi.org/10.2307/378392

Devitt, A. J. (1993). Generalizing about genre: new conceptions of an old concept. College Composition and Communication, 44(4), 573. https://doi.org/10.2307/358391

Devitt, A. J. (2000). Integrating rhetorical and literary theories of genre. College English, 62(6), 696. https://doi.org/10.2307/379009

DiAngelo, R. J. (2016). What does it mean to be white?: Developing white racial literacy (Revised edition). New York: Peter Lang.

Inoue, A. B. (2014). A Grade-Less Writing Course That Focuses on Labor and Assessing. In D. Coxwell-Teague & R. F. Lunsford (Eds.), First-year composition: from theory to practice. Anderson, South Carolina: Parlor Press.

Lorde, A. (1984). Sister outsider: essays and speeches. Trumansburg, NY: Crossing Press.

Miller, C. R. (1984). Genre as social action. Quarterly Journal of Speech, 70(2), 151–167. https://doi.org/10.1080/00335638409383686

Wynne, J. (2008). We don't talk right, you ask him. In L. Delpit & J. K. Dowdy (Eds.), Skin That ¬¬We Speak: Thoughts on Language and Culture in the Classroom. New ¬York: New Press.

An Addiction to Capitalism: A Criticism of Mainstream Environmentalist Rhetoric

Jake Engel[1*]

ABSTRACT— Inspired by research on the politics of sustainability and the rhetoric of the political economy, this paper will analyze and critique the rhetoric of mainstream environmentalism using Marxist theory. Specifically, my goal is to demonstrate and analyze how capitalism has co-opted environmentalism by individualizing and thus commodifying outrage inspired by the justice movement, reflecting Marx's critique of capitalism's malleability in the face of public outrage. As will be demonstrated, Marxist theory warns that this adaptation is only possible so long as the veneration of capitalism remains axiomatic. As such, this analysis is complimented with scholarship on neoliberal ideology to demonstrate how the corporate co-optation of environmentalism has led to what Antonio Gramsci referred to as a "Passive Revolution." In this light, an internationally recognized "solution" known as Cap-and-Trade is also criticized for following a similar path of commodification. As such, I offer a preliminary course of action based on the global research of economist Thomas Piketty and recognize the tangible progress of Representative Ocasio-Cortez and her proposed Green New Deal. I conclude with the need for further research and scholarly collaboration in the ongoing dialogue on the best discourse and praxis to harness a path towards a sustainable future.

KEYWORDS—*environmentalism, Neoliberalism, Marxism, sustainable development, green capitalism, rhetorical criticism, social movements, co-optation*

There is a formidable myth in our political discourse that assumes that greater environmental protection will result from greater wealth. This myth has its roots in the Environmental Kuznets Curve (EKC), a theory developed in 1991 by Princeton economists Gene Grossman and Alan Krueger (Steinberg, 2015, p. 216) positing that as societies grow richer, environmental degradation will lower due to elevated eco-stewardship engendered by entrepreneurial innovation. This theory is wrong (Wanner, 2015, p. 953; Steinberg, 2015, p. 216). It is well established in the scholarship of environmental studies that wealth does not diminish pollution, but aggravates it instead (Clement, 2011, p. 954; Wheelan, 2010, p. 44; Wanner, 2015, p. 953; Stern, 2004, p. 2; Steinberg, 2015, p. 216). Critically, this scholarship illuminates the often-wide gap that exists between the public costs of production (pollution, environmental degradation, etc.) and the private costs of production (investments made in products by producers and consumers). These gaps, also known as externalities, result in market failures because the private transactions that keep an economy afloat worsen the health of society *as a whole* (Wheelan, 2010, p. 44).

While this rebuttal may be empirically established, the ideology packaged in the EKC maintains a tight grip on popular conceptions of "common sense," entrenching itself in the status quo of policy—namely green capitalism. Green capitalism is an institutional framework established through mainstream environmentalism in which consumers are encouraged to "buy green" in the hopes that they are helping the climate in some shape or form. However, their perceived contribution is akin to a drop in the bucket compared to the high levels of fossil fuels

*Corresponding Author: je753@humboldt.edu
[1]Undergraduate Student
Department of Communication
Humboldt State University, Arcata, CA 95521

Engel, J. 2019. An addiction to capitalism: a criticism of mainstream environmentalist rhetoric. *ideaFest Journal: Interdisciplinary Journal of Creative Works & Research from Humboldt State University* 3: 77–85.

used to produce these products. While for-profit corporations continue to expand their enterprises and degrade the environment, they give lip-service to environmentalism in the form of "sustainable packaging" or "donations to wildlife preservation."

To that end, this paper will analyze and critique the rhetoric of mainstream environmentalism using Marxist theory. Specifically, my goal is to demonstrate and analyze how capitalism has co-opted environmentalism through the individualization and commodification of the justice movement, thus reflecting Marx's critique of capitalism's malleability in the face of public opprobrium. As will be shown, Marxist theory warns that such adaptation is only possible so long as the veneration of capitalism remains axiomatic. In doing so, this analysis will also recognize capitalism's topological nature, demonstrating how class oppression maintains its stability through the capitalist transmogrification of justice movements, like environmentalism, into a classist demonstration of privilege. Before said engagements though, it is necessary to trace the origins of contemporary practices in mainstream environmentalism.

According to Nathaniel Rich of the *New York Times*, contemporary environmentalism can be traced to a series of events initiated by Rafe Pomerance, a Washington lobbyist and environmentalist, and James Hansen, former director of NASA's Goddard Institute (Rich, 2018). In 1979, Pomerance organized for world-renowned scientist Gordon MacDonald to meet with President Carter's top scientist, Frank Press (Rich, 2018, 1.). After an ominous presentation of the data and hazards of climate change, Press ordered Jules Charney, the father of modern meteorology, to organize a meeting with the nation's top climate scientists, including James Hansen, to determine whether the information was accurate enough to be presented to President Carter (Rich, 2018, 1.).

Producing what would be known as the Charney Report, these scientists compared highly advanced computer models of weather systems with one another, fiercely debating calculations, which differed by fractions of a percent. For example, according to a model designed by Hansen, if the atmospheric concentration of CO2 doubles from pre-Industrial levels, the world would heat by 4 degrees Celsius (Rich, 2018, 3.). On the other hand, another model suggested only 2 degrees of warming from the same concentration of CO_2. As Rich ominously notes, "The difference between the two predictions, between warming of two degrees and four degrees, was the difference between thinning forests and forests enveloped by desert, between catastrophe and chaos" (Rich, 2018, 3.).

Ultimately, the Charney Report concluded that "... when carbon dioxide [levels] double in 2035 or thereabouts, global temperatures would increase between 1.5 and 4.5 degrees Celsius, with the most likely outcome of three degrees" (Rich, 2018, 3.). For the next four decades and beyond, this report would inform numerous international negotiations and treaties aimed at hindering a changing climate. However, once Ronald Reagan was elected President and neoliberalism planted its seeds in American and international government, leading scientists like Hansen questioned "...whether what had seemed to be the beginning had actually been the end" of their progress (Rich, 2018, 6.).

As one can imagine, this report spelled trouble for Big Business, specifically the Fossil Fuel industry. The world's best scientists had come together to emphatically predict global catastrophe by 2035 if CO2 levels didn't lower. This meant drastic measures would have to be taken on a global scale to hinder CO2 concentration fast enough to secure a healthy planet before time ran out. In defense of profit then, the Fossil Fuel industry had to convince enough people that economic growth and environmental stewardship could complement each other. They had to convince the world of green capitalism.

Green capitalism came to the forefront of environmentalism in 1983 when Norwegian Prime Minister Gro Harlem Brundtland was chosen by the Secretary General of the United Nations to lead the Brundtland Commission, a commission that was charged with overseeing global environmental and economic development (Carruthers, 2001). What is important to note about this commission is that, in contrast to earlier forms of environmentalism which understood "sustainability" as the philosophy and practice of challenging hegemonic institutions of production under the guise of planetary limits to growth, the commission manipulated this narrative to *maintain* these institutions under the direction of market-oriented solutions to eco-degradation through the rhetoric of "sustainable development". Famously, Brundtland declared in the Brundtland Report that "Humanity has the ability to make development sustainable—to ensure that it meets the needs of the present without compromising the ability of future generations to meet their own needs" (Carruthers, 2018, p. 98). While this sounds agreeable, the rhetoric was quite strategic (Measter and Japp 1998). Brundtland's goal was to secure the trust of environmentalists while maintaining the privilege of corporate elitists to operate under vast deregulation and neoliberal ideology; her statement was "friendly" enough to captivate environmentalists while maintaining enough ambiguity to wink at the business community in fidelity. What would follow are decades of neoliberal global policies aimed at fattening the fat cats while degrading the environment in virtually any way that produced profit.

Consequently, such connivance has fomented habits of trickery among several multinational corporations (MNCs). According to Phaedra Pezzullo in *Toxic Tourism: Rhetorics of Pollution, Travel, and Environmental Justice*, for example, "October was designated National Breast Cancer Awareness Month in 1984 by Zeneca, a subsidiary of Imperial Chemical Industries Ltd" (Pezzulo, 2009, p. 108). Initially, then, we were led to praise Zeneca (which later merged to become AstraZeneca) for their leadership in promoting awareness of Breast Cancer. As Pezzulo notes, AstraZeneca "...could be saving women's lives [as] increased awareness motivates a greater number of women to be screened for cancer" (Pezzulo, 2009, p. 110). One of the most insidious characteristics of corporate promotions of social responsibility, however, is what they *don't* convey.

Critically, what AstraZeneca has done is engage in the "co-optation" of a social justice and environmental issue (Pezzulo, 2009, p. 112). Co-optation is a term used by environmentalists to refer to any institution of hegemony that "shares" the promotion of a certain issue with the very disenfranchised groups who are affected by it. While some may view this as a progressive step in corporate behavior, Pezzulo notes that "The problem of corporate public relations is to reposition commodities whose production and consumption may be damaging to the physical environment as 'earth-friendly'" (Pezzulo, 2009, p. 112). For example, the Toxic Links Coalition (TLC) was formed in 1994 in response to AstraZeneca's deceiving rhetoric (Pezzulo, 2009, p. 113). Specifically, TLC has argued that AstraZeneca "...profits first by producing many of the toxins implicated in the breast cancer epidemic and then by selling the drugs used to treat the disease" (Pezzulo, 2009, p. 114). In other words, by poisoning women's bodies and then selling them drugs as therapy, AstraZeneca has positioned itself as both the cause and the "solution" to breast cancer. As a result, TLC wants to "Shift public discourse about breast cancer from promoting mammograms to 'what might be causing breast cancer' or to 'the environmental causes of cancer'" (Pezzulo, 2009, p. 114).

This, of course, is just one example of corporate contradictions in their rhetoric of sustainability. According to Peter Dauvergne in "The Sustainable Story: Exposing Truths, Half-Truths, and Illusions," MNCs are making grandiose claims such as "100 percent recycling; zero waste to landfill; 100 percent sustainable sourcing; 100 percent carbon neutrality" and more (Dauvergne, 2016, p. 391). In the case of Coca-Cola, this MNC has even managed to partner with the prominent non-governmental organization, The World-Wide Fund for Nature (WWF), for Arctic conservation. In lieu of their 2014 deal, if Coca-Cola matched WWF donations of $1 million, WWF agreed to promote Coca-Cola's brand on their website as a sustainable company, "...praising [them] for their generosity" (Dauvergne, 2016, p. 394). In fact, after Coca-Cola labeled its cans with polar bears for a month in 2011, WWF declared that this MNC is "...more important, when it comes to sustainability, than the United Nations" (Dauvergne, 2016, p. 394). Critically then, as MNCs disingenuously boast of their sustainable practices, an insidious marketing strategy called "greenwashing" has supported their rhetoric.

According to Nick Feinstein in "Learning from Past Mistakes: Future Regulation to Prevent Greenwashing," greenwashing is broadly defined as any "...false assertion [or] claim that exaggerates, misdirects, or misleads consumers as to the environmental qualities of a product" (Feinstein, 2016, p. 233). As such, when companies boast of the sustainable foundations of their products, the logic underpinning their rhetoric encourages further engagement with the market as a means to protest that very market. Critically, "...the late 1980s created a new breed of consumer who demanded environmentally responsible products. Almost overnight, green consumerism transformed the niche market for ecologically safe products into a mainstream industry" (Feinstein, 2016, p. 230).

For example, in "Individualization: Plant a Tree, Buy a Bike, Save the World?", Michael Maniates argues that mainstream environmentalism has commodified and individualized climate action by holding consumers accountable for their purchasing habits as opposed to targeting the Fossil Fuel industry for maintaining the status quo of production (Maniates, 2012, p. 34). In other words, rather than encourage open political engagement to challenge the power structures of a class system, mainstream environmentalism has transmogrified the social infrastructure of its movement into a classist product with a price tag. In this way, capitalism maintains its ability to commodify our bodies, our environments, and the needs of both through the continued nourishment of class asymmetries via the privatized and thus privileged access to certain goods deemed "appropriate" by mainstream environmentalism. In this light, it is now necessary to unpack precisely how environmentalism can be "commodified" by examining the nature of a commodity.

According to Karl Marx in *Das Kapital*, Part 1, "A commodity is, in the first place, an object outside us, a thing that by its properties satisfies human wants of some sort or another" (Tucker, 1978, p. 303). As implicated in this definition, the nature of a commodity can be understood by articulating the difference between a "*need*" and a "*want*" and how both are often conflated through social conditioning. A useful reference is Maslow's Hierarchy of Needs (McLeod 2018). This hierarchy is demonstrated in a triangle, like that of a food pyramid. On the very bottom are basic physiological needs like adequate food, shelter, and environment; as the level increases, higher needs like community and self-esteem become central to one's health. Meeting these higher needs can be difficult though, since people vary in how to define, establish, and maintain healthy communities and self-esteem. As a result of this ambiguity, these needs are often left malnourished and are thus more susceptible to systemic translation into culturally conditioned wants such as over-consumption and lifestyles defined by material goods. Critically, this dialectic of desire is evidenced in the environmental rhetoric of sacrifice (Meyer 2010).

Under the guise of limiting consumption to sustain ecosystems, eco-conscious consumers exist on a dialectic between what Todd McGowan refers to as "societies of prohibition" and "societies of commanded enjoyment" (Stravakakis, 2006, p. 100). In societies of prohibition, individuals are asked by their peers and media to engage in asceticism for the health and safety of the community (Stravakakis, 2006, p. 100). In societies of commanded enjoyment, however, the asceticism required by previous acts of sacrifice is "rewarded" through the mediated encouragement of over-consumption (Stravakakis, 2006, p. 100). Accordingly, eco-conscious consumers are found both yearning for moral approval from the rhetoric of sacrifice, and eager to "earn" their visas into societies of commanded enjoyment to extol and engage in cathartic materiality. Conveniently, the connivance of capitalist constructions of desire makes the supposed cathartic nature of said materiality an unreachable state of mind, teasing consumers with the ideology of satisfaction as if by dangling a carrot in front of a working horse (Stravakakis, 2006, p.100). In this way, the politics of consumption are maintained through a "...tripartite nexus connecting economy, desire, and power" (Stravakakis, 2006, p. 100), each category reinforcing a culturally engineered syllogism of obedience. (As Maniates notes, "In our struggle to bridge the gap between our morals and our practices, we stay busy—but busy doing that which we're most familiar and comfortable: consuming our way to a better world") (Maniates, 2012, p. 37).

Accordingly, the rhetoric of mainstream environmentalism can be best understood with reference to the Marxist saying, "First as Tragedy, Then as Farce" (Tucker, 1978, p. 594). In the *Eighteenth Brumaire of Louis Bonaparte*, Marx applied his theory of historical materialism—the idea that material conditions have historically influenced thoughts, behaviors, and outcomes—to illuminate how power was merely recycled by the French Revolution (Tucker, 1978, pp. 600, 601, 606). He noted that after Napoleon Bonaparte inspired the French peasants by helping them overthrow the Monarchy of Louis Phillippe, his nephew, Louis Bonaparte, took advantage of this populist wave and established himself in a farcical Bourgeois republic (Tucker, 1978, pp. 192, 193).

Because French farmers were so poor and alienated from the rest of society, they had no way of representing themselves in the larger system (Tucker, 1978, p. 608). Recognizing a political opportunity to capitalize on a vulnerable population then, Louis Bonaparte persuaded them that he could improve their conditions. What happened, however, was that the farmers became victims of control under his regime, and the corresponding bourgeois class maintained its power. Bonaparte had used his position to milk all the money from their small holdings through taxes to enrich himself and the Bourgeoisie to maintain the status quo of power distribution (Tucker, 1978, pp. 610, 611, 612). Moreover, the term "tragedy" alone is critical in understanding the contemporary context of environmentalism.

This term was coined by the Ancient Greeks to describe a popular genre of theater in which the ending always resulted in the death of the main character. Critically, what made this genre a tragedy was the fact that there was always an inherent quality of the main character which made their death inevitable. Upon reflection of green capitalism then, not only are most corporate claims of sustainable practices "...false, misleading, or unsubstantiated" (Feinstein, 2016, p. 233), but a popular contradiction in environmental studies called the Jevon's Paradox is reinforced (Clement, 2011, p. 954).

The Jevon's Paradox, a theory coined by British economist William Stanley Jevons, notes that if companies innovate their production to require less resource use, then net usage of that resource will actually *increase* (Clement, 2011, p. 954). For example, Jevons observed that "Increases in the efficiency of the production of coal meant greater profits, which would attract more investment, ultimately expanding the scale of coal production" (Clement, 2011, p. 954). Moreover, according to a state-level analysis of carbon emissions in the United States from 1963–1997, "The average carbon intensity declined by about 30%, while the average amount of total CO2 generated by the sample used in this study increased by slightly more than 122%. During the same period, the average increase in total economic product was roughly 242%" (Clement, 2011, p. 954). Accordingly, even if companies reduce their emissions and engage in sustainable practices, the fact that participation in capitalism necessitates endless economic growth means we must do one of two things: regulate corporations or explore a different system of production.

With respect to the first solution, the most popular internationally recognized method of carbon reduction is a program called Cap-and-Trade (Kill, Ozinga, Pavett, et al. 2010). Under this program, most governments of industrialized countries distribute tradeable permits to qualifying polluting businesses (Kill, Ozinga, Pavett, et al., 2010, pp. 29, 30). These permits allow companies to emit carbon up to a limit, or a cap, set by the respective governments of the international agreement. Accordingly, this program follows quintessential market logic: if a company is wealthier and able to innovate, they will have left-over permits with which to trade less innovative companies for money (Kill, Ozinga, Pavett, et al. 2010). The intended result is such that no company is "overregulated" and can instead increase their bottom line through appropriate behavior (Kill, Ozinga, Pavett, et al., 2010, pp. 17, 32; Steinberg pp. 98, 112). Here's the catch: the system lacks rigorous oversight (Kill, Ozinga, Pavett, et al., 2010, pp. 27, 43, 52).

Inevitably, this means wealthy companies are incentivized to deceive regulators, continue polluting at current rates, and profit from poorer businesses via the exchange of "left-over" permits for money (Kill, Ozinga, Pavett, et al., 2010, p. 45). As such, the program of Cap-and-Trade has invented a market wherein businesses have legitimate claims to property rights over their emissions in the form of these permits. Therefore, this internationally practiced regulation has managed to commodify the very air we breathe by asking citizens to pay taxes for access to quality air in the form of a highly ineffective regulation. In this case, of course, the commodity represents a legitimate need, but this need becomes privatized, resulting in a want-based market. The other option (a different system of production) faces a formidable ideological barrier in neoliberalism.

Before examining this option, it would behoove me to explain the ideology of neoliberalism since it has been used throughout this paper. According to David Harvey in "A Brief History of Neoliberalism," "Neoliberalism is in the first instance a theory of political economic practices that proposes that human well-being can best be advanced by liberating individual entrepreneurial freedoms and skills within an institutional framework characterized by strong private property rights, free markets, and free trade" (Harvey, 2005, p. 2). In other words, neoliberalism is an ideological flavor of capitalism that favors deregulated competition, individuality, and grit as the means to achieving a prosperous society. This ideology sounds familiar (even comfortable) because it's the same ideology that has run much of the world since its launch into mainstream practice during the administrations of Margaret Thatcher and Ronald Reagan in the 1980s.

Surely, one might argue, it mustn't be *that* bad if every President since 1979 has incorporated its philosophy into the guiding principles of their administrations; certainly, each President has had the privilege of electoral success to attain such a position in the first place. As Harvey notes though, the ideological power of neoliberalism operates in a highly insidious manner (Harvey, 2005). If enough people have had to be convinced of neoliberalism's merit to maintain its philosophical bearings throughout every Presidential administration since Reagan, then the strategies of persuasion employed must take clandestine, all-encompassing forms to accomplish such mass deception (Harvey, 2005, pp. 39, 40).

In line with Luis Althusser's acute observation of Ideological State Apparatuses (Althusser, 2014, pp. 80, 81), Harvey notes that the ideology of neoliberalism has been infused into everyday rituals and practices, masking itself as "common sense" (Harvey, 2005, pp. 39, 40, 41). We can reflect on Maniates's observation of the individualization of environmentalism to better understand this topology. Individuals no longer need direct persuasion of the merits of sustainable free enterprise before they purchase a new hybrid or other "eco-friendly" products. The ideology of neoliberalism has become so hypostasized through years of persistent marketing strategies, think tank studies, and elite academic cadres, that we've come to embody its philosophy as an unquestioned *performance*. Borrowing Judith Butler's Theory of Performativity, just like gender is an embodied performance ritualized into everyday acts of repetition no longer subject to conscious surveillance (Gunn & Treat, 2005, pp. 161, 163), neoliberalism has tainted all institutions of ideological bearing including education, political parties, media networks, and familial relationships to subsume "rational" decision-making under the parameters of unconscious conditioning (Althusser, 2014, pp. 80, 81, 83; Harvey, 2005, pp. 39, 40, 41). In this way, the barrier to realizing an alternative system of production is found right where pollution started in the first place—at the whim of the owners of polluting businesses (Tucker, 1978, pp. 174, 175).

According to Marx in Part 1 of the German Ideology, the ruling class maintains power through the promotion of its ideology as "...the common interest of all the members of society" (Tucker, 1978, p. 174). For example, Marx suggested that the "Trick of proving the hegemony of the spirit in history is thus confined" to separating those in power from their ideas (Tucker, 1978, p. 175). In other words, so long as we ignore the connection between those in power and the corresponding ideology that complements their power, the hegemony of Bourgeois ideology will remain veiled as that of the people. Reinforcing this hegemony is the discourse of catastrophism (Foust and Murphy, 2009, pp. 153, 154, 161; Yuen, 2012, pp. 19, 20, 32).

In Foust's "Revealing and Reframing Apocalyptic Tragedy in Global Warming," the doomsday rhetoric of environmentalism in mainstream media is problematized as encouraging a sense of inevitability, thereby restricting individual spirit to enact change (Foust, 2009, pp. 152, 153). For example, Foust begins by referencing Al Gore's award-wining documentary, *An Inconvenient Truth*, as a vantage point off which to analyze and critique popular conceptions of the environment through the lens of epideictic rhetoric. In this film, Gore overwhelms the audience with incessant visions of real-life disaster only to conclude that the best strategy to combat climate change is to adopt an individualistic approach and vote with our dollars. As such, Foust critiques this narrative, arguing that by situating disaster as inevitable and its solution as individually oriented, we strip individuals of their sense of collective agency and motive to take more ambitious measures that extend beyond the market (Foust, 2009, pp. 153, 154). As a result, Foust argues, the rhetoric of catastrophe (or apocalypse) mediates individual behavior to remain entrenched in the capitalist framework as if through a placebo effect (Foust, 2009, p. 153).

Moreover, as Yuen notes in "The Politics of Failure Have Failed: The Environmental Movement and Catastrophism," the discourse of catastrophe has an antiquated record in our collective memory, resulting in a sense of "catastrophe fatigue" (Yuen, 2012, p. 20). For example, in the film *2012*, the media industry took advantage of a Mayan prediction that the world would end that year. Once this prophecy was proven false in 2013, the film was merely added to a *rising pile* of blockbuster, catastrophe-oriented media. Accordingly, not only does the rhetoric of catastrophism plague public spirit with fatigue and a consequential loss of perceived agency, but this process inevitably supports capitalism as the best institutional framework to channel a complex mixture of fear, frustration, and said fatigue (Yuen, 2012, p. 32). Specifically, Yuen refers to this process as "catastrophe capitalism" in which markets adapt to environmental criticisms by adjusting their presence to appear more eco-conscious (Yuen, 2012, p. 33). Ironically, just like AstraZeneca was both part of the cause and the solution to breast cancer, capitalism as a socioeconomic system seems to be both crisis-ridden and crisis-dependent (Yuen, 2012, p. 33).

To that end, if the only viable option to prevent a global catastrophe by 2035 is being blocked by the social conditioning of the masses by the few, then perhaps the Industrial Revolution and the ideology that ensued were the beginning of the end of a historic addiction—an addiction to capitalism. Specifically, Thomas Wanner notes that we have entered a stage in this addiction which Italian Marxist Antonio Gramsci coined a "Passive Revolution" (Wanner, 2015, p. 25). A passive revolution, according to Gramsci, is one in which "Counter-hegemonic challenges to the dominant capitalist order are co-opted and neutralized through changes and concessions which re-establish the consent in that order" (Wanner, 2015, p. 25). In other words, we have entered a stage in which the very forces predicted to cause a global demise in the coming decades have adapted to their loudest critics to maintain the status quo through the appearance of progress while simultaneously nourishing class oppression. As part of this adaptation, the ruling class has adopted what economic historian Karl Polanyi termed "fictitious commodities" (Wanner, 2015, p. 25).

According to Polanyi, the term "fictitious commodities" refers to anything that becomes commodified, yet is not meant to be exchanged in the market (hence its fictitious character) (Wanner, 2015, p. 25). Although I have argued that nature, or at least the *idea* of nature, is commodified in the form of hybrids and myriad other "eco" products, the international program of Cap-and-Trade more accurately reflects Polanyi's definition. Under Cap-and-Trade, companies can commodify and privatize healthy air through legitimate claims to property rights of their carbon emissions under the guise of permit use. As such, citizens are then asked by the government to pay taxes to distribute these permits for access to quality air. As discussed though, the oversight and enforcement mechanisms of this program are next to nothing, making the commodification of air free to companies in the position to attain permits. Accordingly, Polanyi mentioned fictitious commodities in his book, *The Great Transformation*, to describe an insidious process by which capitalism slowly subsumes and commodifies everything in society until nothing is barred from producing profit (Wanner, 2015, p. 25).

In this respect, it seems like capitalism, regardless of a neoliberal taint, is proving to be biophysically unsustainable in a culture defined by infinite growth on a planet constrained by limits to resource use. In response to the potential of a withdrawal from such a toxic relationship then, this global addiction has reached the stage of farce—or repetition through irony. As discussed, consumers have become more critical of the lack of eco-stewardship among industry leaders, and are thus demanding more sustainable methods of production. The results, however, have been the corporate greenwashing of products, the reinforcement of class oppression, and the inefficient regulation of industry to deceive consumers into being content with an illusion of progress.

Upon recognition of this dim forecast, there have been several calls to resist "inevitable catastrophe" through the ecological transformation of the political economy (Hawken, Lovins, Lovins 1999; Akuno, Nangwaya 2017; Kurtzleben 2019). Among the most recent and popular is Congresswoman Alexandria Ocasio-Cortez's plan for a Green New Deal (GND) (Kurtzleben 2019). In short, the GND, a bold iteration of Franklin Delano Roosevelt's original New Deal of the 1930s, will entail a stimulus package which pairs populist concerns over class oppression with environmental concerns over a changing climate and environmental degradation by implementing a sweeping set of economic reforms under the guise of a green and just economy. Critically, this vision is echoed by esteemed political journalist Naomi Klein's argument that to properly address the hazards of a changing climate, politicians, cultural activists, and educators alike should bridge the gap between two sets of concerns (a changing climate and class oppression) which, to some, may not have seemed intuitively connected (Klein 2011). The logic behind this convergence is quite sound though. The same global plutocracy responsible for changing the climate is also responsible for the highest levels of wealth inequality in world history. As such, addressing one issue should not have to come at the expense of addressing the other; in fact, doing so would be neglectful policy.

To that end, while details on the proposal are still developing, the general idea is to support a number of public investments to reshape the political and environmental structure of the economy. These investments would include programs like universal healthcare, adopting a 70% marginal income tax on top earners, followed by a timely transition to 100% clean energy (Kurtzleben 2019; Jacobson 2019). In response to this proposal, Senator Sanders recently invited Congresswoman Ocasio-Cortez to a climate summit to discuss the matter. During this time, both acknowledged the unprecedented magnitude of political will and commitment that such a proposal would require (Green 2018). In fact, Ocasio-Cortez declared that properly addressing climate change will be the "Civil Rights movement of our time," noting that any proposal to address the issue must match the gravity of the problem it seeks to solve (Green 2018).

To be sure, the GND will likely be established in the framework of a capitalist market. This will require a set of regulations and other forms of oversight to ensure that green washing dissipates and that Cap-and-Trade will either be conducted with more enforcement, or be replaced by a stronger program. One idea, voiced by renowned French economist Thomas Piketty in his groundbreaking analysis *Capital in the Twenty-First Century*, is a global tax on capital (Piketty pp. 451-455; Mason 2014). While the GND is no doubt a bold step in the compression of inequality, it ignores a sobering detail about the multi-dimensional structure of said inequality—namely the difference between inequality of income and inequality of wealth. Since wealth (or capital) is not accumulated from labor but rather from assets, it is far more mobile and thus remains geographically unconstrained by the inhabitance of its owner.

In short, those who are powerful enough to own capital have the legal privilege to hide that capital across the globe in politically benign territories known as tax havens (the Cayman Islands being amongst the most popular) (Peretti 2016). These tax havens help the 0.01% shield the bulk of their finances from being redistributed through taxes. In this light, Piketty's call for a global tax on capital matches the gravity of the situation in two ways. First, it recognizes the root of inequality and does not mistake earned income for accumulated assets. Second, it addresses the scale of the problem by spanning the globe and not constraining action to geographic limitations. A hefty tax on income is productive, but it provides a national solution to an international crisis. (To be clear, most of the world's wealth does originate from the United States. Esteemed economist Branko Milanovic illustrated this well when he compared various countries' respective share of the global plutocracy (Milanovic, Ch. 1, pp. 36, 37). Moreover, it should be noted that Ocasio-Cortez is a Representative, not a Senator. This is important to mention because Representatives are not responsible for international policy, like treaties. They only engage in domestic politics.)

Therefore, while praising the GND's domestic efforts, it would be wise to supplement its domestic tax on income with a global tax on capital, like that recently proposed by 2020 Democratic Presidential candidate Elizabeth Warren (Yglesias 2019). In doing so, we would need to build on our alliances and call for a multi-facetted treaty to tax capital on a global scale. This would require, among other things, a demonstration that the U.S. honors these alliances at all (one of the most dangerous outcomes of President Trump's blatant disregard of our allies is that international crises like climate change and an ever-increasing global plutocracy will be left untouched due to an acute spike in an international flavor of individualism). Finally, with the revenue accumulated from such a tax, the U.N. should be provided with an independent treasury department responsible for appropriating the funds on regulations and swift transformations to renewable energy around the globe.

As one may note, there has been no mention of any plan to transform the means of production. Moreover, the plan I *have* outlined should not be mistaken to typify an authoritative agenda; much further political analysis is required to comprehend the excruciatingly complex network of actors, values, processes, and institutions involved in addressing such ambitions. In this way, my goal has been to contribute to said dialogue by addressing, discerning, and complicating status quo conceptions of sustainability using critical theory. Perhaps someday, through the interdisciplinary collaboration of scholars, environmental scientists, and political philosophers, a productive course of action can be taken on a global scale to unite the struggles of the working class with those of the planet and future generations.

REFERENCES

Akuno, K., Nangwaya, A. (2017). Jackson rising: The struggle for economic democracy and black self-determination in Jackson, Mississippi. Retrieved April 13, 2019, from: https://darajapress.com/publication/jackson-rising-the-struggle-for-economic-democracy-and-self-determination-in-jackson-mississippi.

Althusser, L. (2014). On the reproduction of capitalism: Ideology and ideological state apparatuses. Verso. (pp. 79–87).

Carruthers, D. (2001). From opposition to orthodoxy: The remaking of sustainable development. Journal of Third World Studies, 18(2), 93–112.

Clement, M. (2011). The Jevons paradox and anthropogenic global warming: A panel analysis of state-level carbon emissions in the United States, 1963–1997. Society and Natural Resources, 24(9), 951–961.

Dauvergne, P. (2016). The sustainability story: Exposing truths, half-truths, and illusions. In New Earth Politics: Essays from the Anthropocene (pp. 387–404). The MIT Press.

Feinstein, N. (2013). Learning from past mistakes: Future regulation to prevent greenwashing. Boston College Environmental Affairs Law Review, 40(1), 229–257.

Foust, C., & O'Shannon Murphy, W. (2009). Revealing and reframing apocalyptic tragedy in global warming discourse. Environmental Communication, 3(2), 151–167.

Green, M. (2018). Ocasio-Cortez: Fighting climate change will be 'the civil rights movement of our generation. Retrieved from: https://thehill.com/policy/energy-environment/419564-ocasio-cortez-fighting-climate-change-will-be-the-civil-rights.

Gunn, J., & Treat, S. (2005). Zombie trouble: A propaedeutic on ideological subjectification and the unconscious. Quarterly Journal of Speech, 91(2), 144–174.

Harvey, D. (2005). A brief history of neoliberalism. Oxford; New York: Oxford University Press. (pp. 79–87).

Hawken, P., Lovins, A., L. H., Lovins (1999). The next industrial revolution. Natural Capitalism - Creating the Next Industrial Revolution. (pp 1–21). Retrieved April 13, 2019, from: http://www.natcap.org/.

Jacobson, L. (2019). Explaining Alexandria Ocasio-Cortez's 70 percent tax idea. Retrieved from: https://www.politifact.com/truth-o-meter/article/2019/jan/08/explaining-alexandria-ocasio-cortezs-70-percent-ta/.

Kill, J., Ozinga, S., Pavett, S., & Wainwright, R. (2010). Trading carbon:

How it works and why it is controversial. FERN. doi: 978-1-906607-10-4.

Klein, N. (2011). Capitalism vs the climate. The Nation. Retrieved from https://www.thenation.com/article/capitalism-vs-climate/.

Kurtzleben, D. (2019). Rep. Alexandria Ocasio-Cortez releases green new deal outline. Retrieved from: https://www.npr.org/2019/02/07/691997301/rep-alexandria-ocasio-cortez-releases-green-new-deal-outline.

Maniates, M. (2001). Individualization: plant a tree, buy a bike, save the world? Global Environmental Politics, 1(3), 31–52.

McLeod, S. (2018). Maslow's hierarchy of needs. Retrieved from: https://www.simplypsychology.org/maslow.html.

Meister R, M., & Japp, P. M. (1998). Sustainable development and the global economy: Rhetorical implications for improving the quality of life. Communication Research, 25(4), 399–421. https://doi.org/10.1177/009365098025004004.

Milanovic, B. (2016). Global inequality: a new approach for the age of globalization. Cambridge, Massachusetts. Retrieved from: https://www.cps.fgv.br/cps/bd/curso/6-Branko-Milanovic-Global-Inequality-Intro-Chapter-1.pdf.

Meyer, M. J. (2010). A democratic politics of sacrifice? The environmental politics of sacrifice (pp. 13–29). Cambridge, Massachusetts: MIT Press.

Peretti, J. (2016). The Cayman Islands – home to 100,00 companies and the £8.50 packet of fish fingers. Retrieved from: https://www.theguardian.com/us-news/2016/jan/18/the-cayman-islands-home-to-100000-companies-and-the-850-packet-of-fish-fingers.

Pezzulo, P. (2009). Toxic tourism: Rhetorics of pollution, travel, and environmental justice. University of Alabama Press. (pp. 1–280).

Piketty, T. (2015). Capital and wealth taxation in the 21st century. Retrieved from: http://piketty.pse.ens.fr/files/Piketty2015NTJ.pdf.

Rich, N. (2018, August 1). Losing earth: the decade we almost stopped climate change. The New York Times. Retrieved from: https://www.nytimes.com/interactive/2018/08/01/magazine/climate-change-losing-earth.html, https://www.nytimes.com/interactive/2018/08/01/magazine/climate-change-losing-earth.html.

Stavrakakis, Y. (n.d.). Objects of consumption, causes of desire: Consumerism and advertising in societies of commanded enjoyment. Gramma: Journal of Theory and Criticism. doi: 2529-1793.

Steinberg, P. (2015). Who rules the earth?: How social rules shape our planet and our lives. New York, NY: Oxford University Press.

Stern, D. The rise and fall of the environmental kuznets curve. Journal of World Development, 2004.

Stirling, A. (2015). Emancipating transformation: from controlling 'the transition' to culturing plural radical progress. In Scoones, I., Leach, M., & Newell, P. (Eds.) The politics of green transformations. Pathways to sustainability. (pp. 54–67). Routledge, London.

Tucker, R. C. (1978). The Marx-Engels reader. New York: Norton.

Wanner, T. (2015). The new 'passive revolution' of the green economy and growth discourse: Maintaining the 'sustainable development' of neoliberal capitalism. New Political Economy, 20(1), 21–41.

Wheelan, C. Naked economics: undressing the dismal science. Norton, 2010.

Yglesias, M. (2019). Taxing the rich is extremely popular. Retrieved from: https://www.vox.com/2019/2/4/18210370/warrenwealth-tax-poll.

Yuen, E. (2012). The politics of failure have failed: The environmental movement and catastrophism. In Lilley, S., McNally, D., Yuen, E., & Davis, J., Catastrophism: The apocalyptic politics of collapse and rebirth. (pp. 15–40). Oakland, Calif.: PM Press.

True Injustice: Cultures of Violence and Stories of Resistance in the New True Crime

Marcos A. Hernandez[1*]

KEYWORDS—*true crime, social justice*

| *Warning: The content of this paper includes topics that may be difficult for some people to confront or discuss.* |

A brief perusal of media coverage on "true crime" today may well leave one convinced, as I was when I first dived deep into it in early 2018, that the genre has experienced something of a cultural renaissance in the last five years. On April 29th, 2019, *Variety* magazine published an article titled, "Inside the True Crime Boom Taking Over Prestige TV." I mention this article not because it is particularly convincing; the writer credits *American Crime Story: OJ Simpson*, the first season of Netflix's true crime anthology series which debuted in 2016, for the recent rise of interest in true crime, even though the resurgence of the genre can be traced back at least to 2014's *Serial*. The article is, however, representative of the broader cultural feeling about true crime shared by a great number of readers and critics today. The title captures what many have perceived to be the two major developments in the genre's recent history that together constitute its renaissance. The first of these is the genre's seemingly rapid rise in popularity as of late, as illustrated by the writer's reference to the so-called "true crime boom," a phrase that has gained currency over the past few years in popular writing about contemporary true crime. The second and arguably more contestable development, suggested by true crime's alleged takeover of "prestige TV," is the general feeling among observers that true crime has become more sophisticated in recent years and has itself acquired a degree of prestige in contemporary society. In this view, offerings such as 2014's *Serial* and 2015's *Making a Murderer* have elevated a genre previously dismissed as "lowbrow" and "popular" to the ranks of high culture. True crime has become, in a word, respectable. For the lifelong student of popular culture and occasional student of cultural studies, three questions naturally arise: are we really in a true crime boom, is it really the case that true crime has gone highbrow, and, if the answer to the first two questions is yes, then why now?

*Corresponding Author: mah123@humboldt.edu
[1]Graduate Student
Department of English
Humboldt State University, Arcata, CA 95521

Hernandez, M. A. 2019. True injustice: cultures of violence and stories of resistance in the new true crime. *ideaFest Journal: Interdisciplinary Journal of Creative Works & Research from Humboldt State University* 3: 86–98.

It would be hard to deny that true crime has experienced a boom in recent years. The "post-*Serial* true crime boom" (as another common collocation has it) saw a revival of interest in the genre. True crime podcasts proliferated in the mid-2010s, from which emerged such fan favorites as *My Favorite Murder* (2016–present) and *S-Town* (2017). Streaming media services have shown no signs of slowing down their true crime output following the success of 2015's *Making a Murderer* and other true crime docuseries, and why would they? A third-party poll found that more than 19 million viewers on average streamed season one of *Making a Murderer* within thirty-five days of its December 2015 release (*Adweek.com*, 2016). Not surprisingly, Netflix is set to release four new true crime Netflix Original Documentaries in June 2019 alone (*Shortlist.com*, 2019). With all the evidence out there, it is getting harder and harder to maintain that true crime is not in the midst of something resembling a boom (a moment, maybe?). Even those who initially balked at the idea of a "true crime boom" have had to concede that something noteworthy is happening. Writing for CrimeReads.com earlier this year, host of the podcast *Criminal Broads* and "Lady Killer" expert Tori Tefler admitted that although the true crime business has always "hummed," "[w]e are clearly in the midst—or at the peak?—of a very specific sort of true crime boom that was midwifed into being with "prestige" true crime like *Serial* and *Making a Murderer* and *The Jinx*." Here, again, the causal link is established between the genre's rise in popularity and its perceived newfound prestige. The underlying logic of this argument seems to run as follows: True crime was previously a lowbrow form of popular art that fed the uneducated public's insatiable appetite for lurid and sensational tales of violence, but recently the genre has become more intellectually-engaging, sophisticated, and respectable in both content and form, and we are witnessing a rise in the genre's popularity *because* of this change in overall quality. This view is problematic (not to mention contradictory) for several reasons, not least because it assumes that a distinction between high and popular culture exists by which cultural products such as the true crime story can be categorized.

In an essay titled "What is Popular Culture?," cultural critic John Storey sketches out five conceptual understandings of the phrase "popular culture" in current general and critical usage. The current praise of "the rise of highbrow true crime" seems to rest upon Storey's second definition, which views popular culture as occupying the subordinate position within a binary structure. In this definition, popular culture is

> the culture which is left over after we have decided what is high culture ... a residual category, there to accommodate cultural texts and practices which fail to meet the required standards to qualify as high culture. In other words, it is a definition of popular culture as inferior culture.
>
> 6

Popular culture is thus defined in negative relation to high culture. In this Arnoldian tradition, high culture encompasses "all the best that has been thought and said" by a society, while popular culture is made up of all works that have been deemed to lack the qualities of high art. This division between high and low culture is used, mainly but not exclusively, by members of the cultural elite to police boundaries between supposedly superior and inferior forms of cultural production. Such distinctions presume that the standards of cultural taste are universal (stable across human experience) as well as trans-historical (fixed for all time). One unfortunate legacy of this perspective was the exclusion of popular art as a worthy site of cultural study.

Cultural critics such as Storey have long rejected the binary structure upon which the division between a high and low culture rests, insisting that the cultural tastes of any day are socially constructed, locally and temporally contingent, and, as a result, always open to contestation. Consequently, there can be no universal or trans-historically fixed criteria by which to categorize cultural products as either good or bad or high or low. This critical perspective has had two lasting legacies. First, this view rendered evaluation of cultural products based on aesthetic quality a highly untenable task. How can you definitively judge whether a work of art is good or bad or high or low when there are no legitimate grounds upon which to base aesthetic judgements? With a void left by the death of aesthetic judgement, cultural critics now shifted their critical gaze to issues of ideology, representation, and the cultural production of meaning in the analysis of culture. Second, by rejecting low/high culture distinctions, cultural critics effectively legitimized popular culture as a site worthy of critical and cultural inquiry. As Barker and Jane explain, "this stance had the great merit of opening up a whole new array of texts for legitimate discussion" (53). In the decades following the emergence of British cultural studies, scholars would take up popular art forms ranging from rock and roll to blackface minstrelsy as legitimate objects of cultural concern.

When looking at contemporary true crime through a cultural studies lens, it becomes clear that something is different about the genre. I would argue that the quality of true crime has changed, but not in the high culture, low culture sense which finds popular expression today in readily available phrases such as "prestige true crime" or "highbrow true crime." Rather, some of the newer entries in the genre suggest that true crime is undergoing something more resembling an ideological shift towards the *left*. True crime has a long history of affirming conservative notions of justice, even when it has been subversive[1]. Yet, more true crime today is beginning to focus on and advocate for those who may have been wronged by the criminal justice system. *Serial* and *Making a Murderer* are indicative of this trend. The true crime storytellers of today, it seems to me, are setting out with other social and political purposes in mind. So, while I agree with the general feeling that the quality of true crime has changed, I do not believe that it has changed in terms of aesthetic quality (i.e. highbrow true crime) but rather in terms of social and political ideology.

Returning to the question of whether we are currently in the midst of a true crime boom, Tefler gives us good reason to curb our enthusiasm about what many have perceived to be the exceptional rise of true crime in recent years. She writes:

> Today, we're in a true crime boom because we're in an everything boom. We are drowning in content these days: with an Xfinity subscription, a Hulu login filched from your kid cousin, and a bit of honest American God-given laziness, you can literally "stream content" until you die. Is there a lot of true crime content floating around these days? Yes, just like there are a lot of music videos and Gossip Girl knock-offs and stand-up specials floating around these days, not to mention so many original Netflix shows that they've been mocked on SNL. And I'm not even going to mention the podcasts again.

Tefler makes a convincing case that perhaps the true crime boom is not so exceptional, after all. In the internet age, anyone with access can find endless entertainment to fill whatever their heart desires. If you love cooking shows, a quick internet browse will pull up tens (if not hundreds) of shows to watch, multiple social media fan groups to join, and any number of celebrity chefs to follow. We have seen the proliferation of not only true crime content, but *all* content. Nevertheless, we are still very much in a true crime moment, and just because everything else is booming, too, does not mean that we are prohibited from looking at the particular forces that have propelled the recent rise of true crime.

If we accept that we are in a true crime moment and that, as I argue, the social and political motives of true crime have started to change, the question then becomes: why now? Tefler's point that we are in an "everything-boom" provides a general starting point, but what is it that has fueled true crime's boom in particular? What circumstances have led the genre to shift ideologically? On the rise of what she dubbed, in a 2015 article for *The Atlantic*, "The New True Crime," Lenika Cruz offers one take when she writes that "[n]ew forces—improved technology, new media, and less trust in institutions—have helped shaped true crime into a truly modern form." There is certainly truth to her claims. As both Tefler and Cruz would agree, increased connectivity and new media forms have undeniably helped shape the true crime moment in which we are currently living. This fact is perhaps no more apparent than when looking at the *Serial* phenomenon. The unprecedented success of *Serial*'s first season demonstrated that the podcast, as both a technology and a new media format, is singularly effective for an incisive study of the violent crime, its precipitating events, and its juridical aftermath. With *Serial*, the podcast's viral capability, combined with the creative affordances of the format, came together to offer an infinitely shareable, easily digestible, and thrilling way to consume true crime stories. To put true crime's natural affinity with the podcast in perspective, *Serial* currently holds the distinction of being the fastest podcast to reach 5 million downloads and streams in iTunes history (*The Guardian*, 2014) and remains the most popular podcast to date. In the post-*Serial* true crime boom that followed, the podcast experienced a stellar rise as the medium of choice for both true crime storytellers and listeners alike. Similarly, streaming media services have widened the genre's reach by offering audiences around the world binge-able and (for the most part) smartly-conceived stories of crime, such as *Making a Murderer* (2015–present) and HBO's *The Jinx: The Life and Deaths of Robert Durst* (2015). Both shows were met with general acclaim and have received

1 Laura Browder (2010) writes that true crime books often "uphold conservative values—policemen are heroes, criminals are punished, sometimes by death ... some of the most successful writers who followed [Capote] have framed their stories as morality tales ... Yet true crime books are also subversive, in that they tend to question the very foundations of patriarchal culture" (126).

multiple accolades. It's also worth noting that while new technology and new media have perhaps played the largest role in re-invigorating the genre, true crime has also continued to be immensely popular in print form, as made clear by the success of recent books like David Grann's *Killers of the Flower Moon: The Osage Murders and the Birth of the FBI* (2017) and Michelle McNamara's *I'll Be Gone in the Dark: One Woman's Search for the Golden State Killer* (2018). A film adaptation of *Killers* is currently in development with Martin Scorsese and Leonardo DiCaprio at the helm, if that's any indicator of the genre's continued relevance in U.S. culture.

Social media has also contributed to the true crime moment by allowing ever greater participation among showrunners, investigative journalists, local and federal authorities involved in particular cases, the families and friends of victims, and fans-turned-amateur-sleuths. In some cases, such "crowdsolving," as it has sometimes been called, has led to breakthroughs in investigations or court trials. News articles with the headlines reading something like "How a true crime podcast helped solve a 30-something-year old cold case" or "How Social Media Helped Solve a Murder" have become a regular occurrence. Just recently, a story broke reporting how listeners of journalist Hedley Thomas' podcast "The Teacher's Pet" helped solve a 36-year-old cold case in Australia (*NPR*, 2018). Though it remains to some degree a fringe internet sub-culture, web-sleuthing is alive and well during the true crime boom, and social media has played a central role in maintaining this subculture. Furthermore, outlets such as Facebook and Reddit have created new digital spaces for fans to trade theories, advocate for justice, provide tips and information, and fuel the fandom. A look at the numbers is revealing. As of this writing, 527,758 people have liked the "Making a Murderer" Facebook page; 446,558 have liked "Serial." The subreddit r/True Crime has 87,822 subscribers, r/serialpodcast has 60,885, and r/makingamurderer has 73,800. These spaces have acted as incubators for true crime. As Cruz explains, "[s]ocial media supports the quick ascendance of particular stories, allowing a grassroots energy to buoy otherwise niche cases to the top of the trending list." In effect, social media has made true crime at once more participatory and more viral.

Cruz's claim that a mistrust of institutions has helped shape contemporary true crime helps us to move beyond the tech and media-based arguments often given as reasons for the current true crime boom to look at how wider cultural shifts in the arena of politics and ideology have influenced the genre. People have become more skeptical about the institutions that are meant to serve and protect them, and their interest in such shows as *Serial* and *Making a Murderer*, which both focus on the possibility of wrongful conviction and imprisonment, reflect the profound skepticism of institutions that has emerged in contemporary society. What's more, contemporary true crime has on several occasions demonstrated its ability to affect change in the fortunes of their subjects. Notably, *Serial*'s investigation into the murder of Hae Lee Min led to a retrial for Adnan Syed, Min's boyfriend who was tried and convicted for her murder. Just recently, Steven Avery of *Making a Murderer* fame won an appeal and will have his case re-examined in a Wisconsin state court (The Guardian, 2019). The abundance of similar examples has led some to claim that true crime has started to shift towards a genre of advocacy. Cruz speculates, "These long-percolating cultural shifts hint at what true crime's future could look like: less straight entertainment and more advocacy journalism, if not in style, then at least in consequence" (2015).

I am inclined to agree with Cruz that the shifting orientations in contemporary true crime reflect a parallel shifting in cultural attitudes. These last five years have found Americans in a strange cultural moment. The #blacklivesmatter movement has forced the issues of police brutality and mass incarceration back into the national conversation. The unsuccessful fight against the construction of the Dakota Access Pipeline in 2016 re-opened many old wounds for Native Americans. The #metoo movement has revealed the pervasiveness of rape culture around the world. In the U.S., the rising right-wing reactionary movement against these civil resistance movements has found its ultimate spokesperson in Donald Trump. Given this social and political climate, many Americans have understandably become less confident in themselves, their neighbors, their leaders, and more importantly, their institutions. This mistrust has manifested itself in American cultural products. While it used to be that analyses of systemic injustice were limited to college classrooms and other pockets of the intelligentsia, now even our "mainstream" cultural products are beginning to engage with the political realities of race, sex, class, gender, and environment. Given true crime's chosen subject matter, the genre is a natural fit for such critical examinations, and recent offerings such as Ava DuVernay's *When They See Us* suggest that true crime is increasingly moving in this direction. In short, the new true crime as advocacy journalism has emerged partly due to a critical consciousness *gone mainstream* in contemporary U.S. culture.

To be sure, these newer entries into the true crime canon are still generally recognizable as exactly that: true crime. They still rely on the old conventions: the classic whodunit narrative, the return to the scene of the crime, the cliffhanger, the courtroom procedural drama, and the dramatic re-enactment. But there has been an unmistakable shift in tone and purpose in the new true crime. The subjects have changed. "The result," Cruz concludes, "is a genre that's still indebted to decades-old conventions, but also one that has found renewed relevance and won a new generation of fans by going beyond the usual grisly sensationalism."

Of course, true crime hasn't always received such praise. In a 1991 essay, Jack Miles concluded that "most 'true crime' is so entirely without an agenda that there is little to discuss: There is only a questionable entertainment to promote" (p. 64). In his view, the genre serves no well-meaning purpose; it is pornographic, shameless entertainment that profits from victims' suffering, re-traumatizes secondary victims, or mythologizes the perpetrator, and that, often enough, manages to accomplish some combination of the above. True crime is no stranger to criticism or controversy, and it should come as no surprise that the current praise of true crime has been far from universal. Since the beginning of the true crime boom in the mid-2010s, we have seen a proliferation of think-pieces rearticulating for the modern consumer of these tales the moral and ethical dangers perceived to be inherent to the genre. These articles have tended to echo past criticisms of true crime. True crime sensationalizes tragedy. True crime teaches politics of fear and paranoia. Many have pointed out, as Miles did in 1991, the genre's "human cost," that is, its potential to re-harm those who were close to the victim of a crime. No matter how carefully researched or reported, they say, true crime can have unintended consequences for secondary victims, often due to reasons outside of the storyteller's control. This potential for additional harm is an even more pressing issue in our current true crime moment, when social and visual medias have fundamentally changed the ways that audiences can interact with the crime story and its subjects. As Houpt concludes, "[h]arm as a byproduct of entertainment may be the true cost of true crime." The journalist points to the negative impact that the buzz surrounding *Serial* had on Hae Lee Min's family as supporting evidence for this claim.[2] There are, of course, other examples. Another common criticism holds that true crime has a tendency to mythologize and sometimes even romanticize the perpetrators of horrific crimes. Even when the text's orientation towards the criminal is that of clear and unmistakable condemnation, readers can still misappropriate the text. One need only look at the cultural mythos built up around Charles Manson and The Family left by the many books that have been written about the subject, most notably *Helter Skelter*, to see this kind of reader-driven mythologizing at play.[3] Interestingly, many critics who claim woke ethics have repeated this critique of true crime. According to journalist Laura Bogart, the true crime story, particularly the one that is centered on the *serial* killer, tends to promote the trope of the "Broody White Male Anti-hero" at the same time that it diminishes the victims, typically women, but also members of the queer community and people of color. Such a focus on the perpetrator, Bogart adds, can "devalue the victims, and, more broadly, 'set a theme' that certain marginalized groups are 'disposable'" (The Week, 2018).

To some unquantifiable extent, I must concede that these critics have a point. Some true crime, even today, still peddles in the tired tropes and the old sensationalism.

2 In a 2015 post on r/serialpodcast, Hae Lee Min's brother blasted fans of *Serial*, writing, "When I found out there was a subreddit for this, I had to do AMA for reddit community. But sorry I won't be answering any questions because...TO ME ITS REAL LIFE. To you listeners, its another murder mystery, crime drama, another episode of CSI. You weren't there to see your mom crying every night, having a heart attck when she got the new that the body was found, and going to court almost everyday for a year seeing your mom weeping, crying, and fainting. You don't know what we went through. Especially to those who are demanding our family response and having a meetup... you guys are disgusting. SHame on you. I pray that you don't have to go through what we went through and have your story blasted to 5mil listeners."

3 I am reminded of a friend of mine from high school who years ago expressed their infatuation with Charles Manson after reading Helter Skelter. They loved Manson's songs (he was an aspiring musician) and affectionately called him "my baby."

It can be opportunistic and exploitative, and there is always the potential to cause additional harm. In terms of representation, some true crime does, directly or indirectly, mythologize the murderer, romanticize the rapist, or both, often at the expense of the victim(s). For example, the tendency for writers and readers alike to promote the "Broody White Male Antihero" at the expense of others can be seen in contemporary treatments of Ted Bundy. Yet, some writers of true crime have begun to find ways to avoid the ethical pitfalls and resist the problematic tropes that have turned many people against the genre. Some have done this by taking on historical crimes so that the relationship between subjects and spectators is distanced to some degree. Others have placed victims' experiences at the center of their stories, or, better, have let the victims tell their own stories. The result is a true crime that recognizes its troubled history and actively seeks to mitigate any potential harm that may be caused by the stories it investigates. In this objective the writers of true crime are not always successful, but since the genre is not going away anytime soon, any attempt at minimizing harm and resisting stereotypes is, in my view, certainly welcome.

Cruz and others (Rowen 2017) have identified the fact that true crime has shifted towards advocacy journalism by advocating for the wrongfully imprisoned and by so doing exposing flaws in the criminal justice system. I would take this argument one step further by suggesting that some of true crime's newer entries have even started to advocate for social justice. Here is the difference as I see it. True crime as advocacy text advocates for individuals wronged by the criminal justice system: the Adnan Syeds, Steven Averys, and Brendan Dasseys of the world. Conversely, true crime as social justice text advocates for marginalized groups of people, for entire communities who still continue to face cultural, legal, and economic injustice today. True crime as social justice text is therefore distinguished from the former in its focus on justice for *groups* rather than *individuals*. Accordingly, these true crime texts tend to focus on crimes rooted in racist or misogynistic ideologies. So, while I must concede that true crime has the potential to cause harm and contribute to problematic representations of both dominant and marginalized groups, I believe that such an argument is an unfair assessment of the genre as a whole, and I think that the promising trend towards "true crime as social justice text" supports my view.

To further my case that true crime has shifted towards a genre of social justice, this paper reviews four contemporary examples of what I call true injustice, an emerging subgenre of true crime that investigates the myriad individual, social, cultural, economic, and legal failings surrounding a specific crime or series of crimes. I am especially interested in works that deal with injustice against whole groups of people, what I refer to as true crime as social justice, since these texts are illustrative of the emerging social justice orientations that could become a defining feature of the new true crime. I look at three books, David Grann's *Killers of the Flower Moon: The Osage Murders and the Birth of the FBI* (2017), Timothy B. Tyson's *The Blood of Emmett Till* (2017), and Jon Krakauer's *Missoula: Rape and the Justice System in a College Town* (2015). I also consider one example from new media, Netflix's *Examen de Conciencia* (2018). In all of these texts, the specific crime (and in some cases, crimes) under investigation is mapped against a complex web of oppressive social, cultural, legal, and economic discourses and practices—in sum, a culture of violence—that, in effect, rationalizes acts of domestic terror against particular groups of people. *The Blood of Emmett Till* reexamines the infamous 1955 kidnapping and late-model lynching of a young black boy in Mississippi to reveal the violent workings of race in the late-Jim Crow American South. In *Killers of the Flower Moon*, the disappearance of a young Osage Indian woman and the later discovery of two bodies opens a chilling book-length investigation into the mass murder of at least twenty (and perhaps more than a hundred) wealthy Osage over oil money in 1920's Osage County, Oklahoma. In *Missoula*, Krakauer examines a recent spate of sex crimes that took place in a small university town in Montana, and though the book doesn't contextualize these crimes within a larger discussion about rape culture in the United States, I argue that it can still be read as a local case study of rape culture. *Examen*, a foreign language Netflix Original Documentary based in Spain, offers a devastating but necessary glimpse into the widespread culture of sexual abuse within the Roman Catholic Church. The effect of each of these texts is the same. In each of these examples, the authors use the true crime story to expose pervasive cultures of violence directed at marginalized groups of people.

For all the focus on mass injustice, however, these texts do not only tell narratives from the rhetorical position of victimhood. In the course of exposing the culture of violence surrounding a particular crime or series of criminal acts, the writers highlight individual and collective acts of bravery. They tell stories of resistance. We learn how Mammie Till Bradley's courage and political maneuvering during a time of terrible grief effectively mobilized the Civil Rights movement. Nearly one hundred years after the Osage "Reign of Terror," we find that the Osage have managed to maintain a sense of tradition and community, almost as if in spite of the mass local and national conspiracy to erase them and their culture. Faced with fear, self-doubt, and a misogynistic culture that is always quicker to place blame on women rather than the accused, five women find the courage to confront their rapists and seek justice. A survivor joins others in breaking the silence about a history of mass sexual abuse in the Catholic Church in Spain.

My reading of these texts will draw attention to the social justice orientations that are an emerging feature of the new true crime. True crime as social justice text maps the logics and legacies of violence and shows how people have resisted such violence. Of course, we must be careful, for not all true crime is created with the interests of social justice (or even individual advocacy) in mind. For this reason, I encourage consumers of the new true crime to be vigilant about the types of true crime they read, hear, or watch, for the simple reason that, like every genre, true crime is made up of the good, the bad, and the ugly.

CULTURES OF RACIST VIOLENCE: *THE BLOOD OF EMMETT TILL* AND *KILLERS OF THE FLOWER MOON*—The first two books I review in this paper both take up historical race crimes to examine cultures of racist violence in the United States. That is to say, both texts demonstrate how a combination of racist social, cultural, legal, and economic discursive practices resulted in a logic of violence that rationalized the murder of racially-marginalized members of society. In *The Blood of Emmett Till*, Tyson explores how deeply-rooted fears of racial integration across all scales of social life in the late-Jim Crow American South created an atmosphere of fear and violence that made it possible for two men to kidnap and murder a young black boy from Chicago and get away with it. In *Killers of the Flower Moon*, Grann traces the history of oppressive cultural, legal, and economic injustices against the Osage that led to the murder of at least twenty (and perhaps hundreds) of Osage Indians in 1920s Oklahoma. These texts illustrate true crime's ability to shift away from sensational tales of murder towards more critical examinations of a history of racial injustice in the United States. Furthermore, though both texts are historical, I argue that they function as social justice texts in that they ask us, directly and indirectly, to draw parallels between the violent histories they tell and the injustices of today.

The Blood of Emmett Till is nominally about the most notorious hate crime in American history, the kidnapping and late-model lynching of Emmett Till in 1955. The story is so well-known that it hardly needs retelling, but I'll describe it briefly. In 1955, Emmett Till, a young black boy from Chicago staying with family in Money, Mississippi, walks into a local market and allegedly makes a flirtatious comment to the store shopkeeper, a young white woman named Carolyn Bryant. Later, Roy Bryant, Carolyn's husband, hears of this violation of the race-sex taboo. Enraged by Till's affront, Bryant, his half-brother, and a group of unwilling black conspirators kidnap, torture, and eventually murder Emmett Till. Four days after the incident at the market, Till's body is found floating just above water in the Tallahatchie River. The book reads as true crime, containing all the markers of the genre, including a return to the scene of the crime, an exclusive interview with Carolyn Bryant, a dramatic retelling of the court trial, and later in the book, a textual re-enactment of the crime that feels like something out of an episode on Investigation Discovery. But the book is also an incisive examination of race relations in the United States at a critical juncture in the country's history.

Through a seamless blend of true crime storytelling and detailed social history, Tyson explains how a public fear of racial integration, rooted in the twin white supremacist ideologies of biological purity and cultural superiority, created a general atmosphere of racist violence against black Americans in the Jim Crow American South. These discourses were articulated and practiced through a network of racist representational, cultural, legal, and economic codes, such as hierarchical social rules and rituals, discriminatory housing practices, segregated schools, and miscegenation laws. Socially, African Americans were expected to behave a certain way around white people; in the unfortunate event that they stepped out line, they were often severely punished. Legally, black Americans in Mississippi were denied the vote, and their attempts to win the vote were actively suppressed through economic reprisals. When that didn't work, threats of physical violence soon followed. Discriminatory housing practices effectively relegated black Americans to unsafe, unsanitary, and dangerous living conditions. They were denied equal employment opportunities. By mapping this culture of racist violence, Tyson ultimately shows his reader how inflammatory discourses of race worked in the 1950s American South to create social conditions that would come to justify the murder of a young black boy from Chicago. Despite the overwhelming evidence that Bryant and Milam were guilty of kidnapping and murdering Till, both men were acquitted by an all-white jury in September 1955. As Tyson explains, the jury was unwilling to convict the men "because a black boy had insulted a white woman, and therefore her kinsman could not be blamed for killing him.14" (180). It is therefore not an overstatement to say that a culture of racist violence in the late-Jim Crow American South quite literally justified the murder the Emmet Till.

In a similar blend of true crime and social history, *Killers of the Flower Moon* begins in 1921 with the disappearance of Anna Brown, a young Osage woman, from her home in Gray Horse, Oklahoma. A week later, the body of Charles Whitehorn, another local Osage, is found near town with a bullet hole between his eyes. Soon after, Anna Brown's body is finally discovered, and we learn that she, too, was the victim of homicide. David Grann takes up these two crimes as a starting point for examining what one federal agent involved with the ensuing case referred to as a "culture of killing" (308) in 1920's Osage Country and what became known in the intervening years as the "Osage Reign of Terror." In the years following the discovery of Brown and Whitehorn's bodies, tens, if not hundreds, of Osage would either be found murdered or would die on the reservation under mysterious circumstances. In the chapters that follow, Grann recounts how the newly created FBI's investigation into the Osage murders eventually uncovered a mass plot by white Americans to murder Osage Indians for the oil money tied to their headrights. To tell this story, Grann employs all the hallmarks of classic true crime, including the cat and mouse chase between law enforcement and criminals, the whodunit narrative, the chapter cliffhangers, and the courtroom drama. But Killers, as social justice text, is also a disturbing glimpse into the long history of exploitation of and violence against Native Americans in the United States.

The framing true crime narrative of Killers is complemented by a history of the Osage in the United States. When the Osage were forced out of their home in Kansas in the early 1870s, they decided to settle in what would later be named the State of Oklahoma, over lands which white Americans at the time believed to be resource dry. Later, the Osage, who had discovered several small oil deposits in the years following settlement, shrewdly negotiated mineral rights over their new land when coming to an agreement with the United States over the terms of allotment. The Allotment Act passed in 1906 granted every Osage member on the tribal roll a headright, which in turn granted them a share in this mineral trust. The act further stipulated that headrights could be inherited but not sold, and further, that inheritances could be passed down to both Osage and non-Osage persons. The reservation was later discovered to be sitting on top of some of the largest oil deposits in the United States, and the Osage became prosperous due to the country's insatiable thirst for oil. Leases were signed and royalties paid, and by the 1920s, the Osage were the richest people per capita in the United States.

Naturally, many white Americans could not fathom the idea of a "red millionaire" (7), and their growing resentment was expressed through a combination of racist cultural, legal, and economic discourses and practices ("Indian Business," as it was referred to by white Americans at the time) that in their total relations worked to undermine Osage wealth and self-determination. The press would regularly stoke white anxieties by telling wildly embroidered tales about the Osage's supposed excesses. Local business owners would sell Osage various goods and services at exorbitantly higher prices than their white counterparts. The worst of all these practices was the federally-imposed policy of guardianship, passed by the Congress in 1921, whereby any tribe member who was half or more Osage, regardless of age, sex, or social status, was placed under the economic guardianship of a prominent white neighbor. This paternalistic policy essentially made room for further economic exploitation. Grann describes an instance in which a guardian purchased a car for less than five hundred dollars and sold it back to his appointed Osage member for more than a thousand. And in the case that an Osage member died, their guardian would inherit the deceased's headright. Within such a pervasively racist social system, it was only a matter of time before some white Americans realized that the only thing standing between them and the Osage's oil money were the Osage. And it was in a culture that actively de-valued Osage lives that this con could turn into a full-fledged conspiracy. Eventually, federal agents and local Osage involved in the investigation uncovered a mass plot, led by a loose network of white Americans including doctors, lawyers, and criminals-for-hire, to kill Osage Indians so that their guardians could inherit the headrights of the deceased.

CULTURES OF SEXUAL VIOLENCE: *MISSOULA* AND *EXAMEN DE CONCIENCIA*—The following book and docuseries I cover in this paper take up sex crimes in the United States and abroad. In *Missoula*, Krakauer examines a series of campus rapes that occurred in just the last decade in the small university town of Missoula, Montana, home to the University of Montana and The Grizzlies, UM's beloved football team. *Examen de Conciencia* (Examination of Conscience) takes us to Spain, where numerous men tell their devastating stories of childhood sexual abuse at the hands of Catholic teachers and priests. Both *Missoula* and *Examen* reveal the various ways that individuals, local communities, and, more importantly, institutions, fail survivors of sexual violence through victim-blaming, lack of support, and active silencing. Furthermore, while both the book and the docuseries focus on localized instances of sexual violence, they directly and indirectly invite the reader to view these investigations as case studies of a wider global culture of rape.

Missoula is an informative, yet deeply disturbing, look into campus and acquaintance rape. Krakauer chronicles the experiences of five women before their rape as well as how they each deal with its aftermath: their previous relationship with the perpetrator; their feelings of fear, anxiety, and self-doubt after the event; their courageous decision to report their rapes and seek justice in either the university or criminal courts; and the university's, local community's and law enforcement's failures to adequately respond to their accusations of rape. The book employs many of the conventions of true crime but dispenses with others. For example, we are still audience to the courtroom dramas, interviews with involved persons, and textual re-enactments of the crime. However, because Krakauer chooses to believe his subjects, there is no need for the whodunit narrative.

One particularly horrifying aspect of rape culture that is addressed in the book is how local communities and institutions often fail victims of sexual violence by way of victim-blaming. Victim blaming refers to the tendency to assign partial or complete blame to the victim of sexual assault. Krakauer describes numerous instances of victim-blaming in the public's response to accusations of rape and subsequent arrests. For example, when one of The Grizzlies' star players, Beau Donaldson, is arrested for the rape of one of the five women Krakauer follows, one internet forum member is quoted as saying: "First off, chicks exaggerate on rape. Second off, she could sucked his dick and still got rape just because she said she didn't want it later on … And a lot of people lie" (p. 54). Not only does the forum member assign partial blame for the rape on the accuser ("she could sucked his dick and still got raped because she said she didn't want it later on"), but they also perpetuate the "myth of false accusation" ("chicks exaggerate rape … a lot of people lie"). As Krakauer notes, studies show that only 2% to 5% of victims make false accusations. If we accept this statistic, that means that 95–98% of the time, the accusers are telling the truth. Yet, as Krakauer shows, local community members and institutions are more likely to side with the accused. This is far from the only instance of victim-blaming and victim-doubting cited in the book.

Examen de Conciencia is a three-part docuseries that brings to light decades (if not centuries) of sexual abuse in the Catholic Church in Spain. This is the framing narrative: in the first episode, we are introduced to survivor and activist Miguel Hurtado, who is contemplating going public about the sexual abuse he endured when he was a young boy in the church. In a moving monologue at the end of the episode, he lets his audience know that he has decided to go public. The second episode follows Hurtado as he mentally and emotionally prepares for the difficult road ahead. To prepare, Hurtado meets with several other survivors who have gone public to hear about their experience. Finally, in the third episode, Hurtado tells his story. The whole series, we discover, is the public accusation. And this is only the framing narrative. Hurtado's narrative is interspersed with social history and other survivor narratives. There is even a deeply unsettling interview with a confessed abuser.

The series illustrates how communities and local institutions fail to support survivors who have come forward to report sexual abuse. The series also shows how the church and its supporters go through great pains to actively silence accusers through such practices as bribery, threat, and non-participation. At the outset of episode one, Hurtado reflects:

> And one thing I've really noticed is the lack of support from the Catholic Church. From government institutions. And, naturally, I wonder: if victims knew what was going to happen, would they have thought twice and decided not to speak out? Or, despite all the difficulties, is it still worth it?

This lack of support operates in several ways. When victims come forward to report their abuse, the church's typical response has been to deal with the offense internally. In most cases, this means that the accused are suspended for a period of time and/or relocated to another parish. Because of the church's relative autonomy, local governments rarely get involved. Even the local community rallies to protect the church. In the first episode, one of the survivors describes how, when he came forward publicly to accuse a church schoolteacher of abuse, members of the PTA descended upon the church and formed a circle around the building to signal their support for the teacher and church.

STORIES OF RESISTANCE—These texts deal with horrific crimes. In some cases, these crimes have been committed by people with magnetic personalities. Easy as it would be for the authors to focus on the perpetrators or narrate from positions of victimhood, the books, in fact, do the opposite: they use the true crime story to champion victims and tell stories of hope and resistance. These stories, I argue, function as social justice texts insofar that they illustrate how marginalized groups have resisted individual, social, cultural, legal, and economic oppression in the recent or not-so-distant past.

These acts of resistance can be momentous in their effects. In *The Blood of Emmett Till*, Tyson makes clear that Mamie Bradley was not only brave but politically astute in her response to her son's murder. Tyson cites Mamie's decision to reach out to black newspapers upon first hearing that Emmett was missing as evidence for this claim. About Mamie's decision to involve the black press, Tyson writes:

> Thanks to Mamie, Chicago's newspapers, radio, and television were already starting to cover the lynching. A TV news bulletin even interrupted I Love Lucy to report the discovery of the body. Now word spread that Emmett Till's body was coming home to Chicago. Mamie now envisioned God's purpose for her life—and for her son's life: "I took the privacy of my grief and turned it into a public issue, a political issue, one which set in motion the dynamic force what ultimately led to a generation of social and legal progress for this country."32 Unlike any of the white newspapers, soon after Till's lynching the Pittsburgh Courier predicted that his mother's "agonized cry" might well become "the opening gun in a war on Dixie, which can reverberate around the world."33 Activists across the country hoped and believed that this tragedy might be the wellspring of positive change. Mamie had ensured that to her mother's cry would now be added the mute accusation of Emmett's body.
>
> 66

Tyson then shows how Mamie Till Bradley's subsequent decision to hold a public, open-casket funeral, her eloquent and composed testimony during the ensuing trial, and her activism in the years following the acquittal, forced the entire nation to confront the horrors of Jim Crow racism and effectively mobilized the Civil Rights Movement.

These acts of resistance can be small and personal. Miguel reflects on his own recovery at the end of episode one of *Examen*:

> Over the years, I've been through significant personal development. At first, I was a victim. My attitude was passive. I'd lost control over

> my life. And I think that I started to become a survivor when I was able to accept my experience and go to therapy. And, finally, I've taken the next step of becoming an activist.
>
> S1:E1 (49:45–50:15)

Hurtado's personal journey from victim to survivor and finally to activist suggests that confronting one's trauma is itself an act of resistance. Seeking help is often the first step. Krakauer describes a similar case to make this point at the end of his book: "Counseling from a skilled therapist can certainly help. And so can speaking the truth about the unspeakable nature of the harm. By such means, Laura Summers managed to regain her equilibrium and find a measure of peace" (380).

In the third and final part of *Killers of the Flower Moon*, Grann recounts his two visits to the Osage Reservation in 2013 and 2015. While there, he discovered that the Osage people have managed, in small ways, to maintain a sense of community and hold onto their cultural traditions, despite the mass conspiracy one hundred years ago to erase them and their culture in the pursuit of riches. Grann describes his experience attending one of the Osages' ceremonial dances, which are held over several weekends every June:

> These dances—which take place, at different times, in Hominy, Pawhuska, and Gray Horse, three areas where the Osage first settled when they came to the reservation, in the 1870s—help preserve fading traditions and bind the community together.
>
> 265

Other moments during Grann's trip reveal the Osage's resilience. He visits the Osage Nation Museum in Pawhuska, where he finds members of the Osage working to keep the history of their people from being forgotten. While driving to Gray Horse, Grann recalls his surprise at seeing bison—which had been systematically exterminated by white Americans in the early 20th century to deprive Native Americans of one of their main food and craft resources—roaming the plains. Grann later learns that bison had been reintroduced by conservationists some years earlier and were now once again under Osage stewardship.

At another level, these texts directly or indirectly encourage us to trace the logics and legacies of violence through history and/or across locations. For example, Blood directly invites readers to draw connections between the violent history it tells and the injustices many black Americans face today. Racist violence against blacks didn't cease to exist then, despite all the social and political gains made by the Civil Rights movement. At the outset of *The Blood of Emmett Till*, Tyson recalls a similar story from his own past of racially-motivated murder in the South, fifteen years after the shocking events of 1955:

> I knew the painful territory well because when I was eleven years old in a small tobacco market town of Oxford, North Carolina, a friend's father and brothers beat and shot a young black man to death. His name was Henry Marrow, and the events leading up to his death had something in common with Till's.
>
> 2

This story serves as the prelude to a later discussion in the book about the inter-generationality of violence against black men in the United States, one that Tyson traces from the time of Emmett Till to the current day. Cultures of racist violence against black people persist today in both overt and subtle forms. Later, in a chapter titled, "The Children of Emmett Till," Tyson encourages his readers to think about how, as he puts it, "America is still killing Emmett Till" (214). He mentions the 2014 shooting death of Michael Brown in Ferguson, Missouri by a white police officer, the 2014 slaughter of nine churchgoers in Charleston, South Carolina by a white supremacist, and the lesser known murder of James Craig in 2011, who was beaten and killed by a group of suburban white teenagers in Jackson, Mississippi, as examples of the overt brutality still experienced by black people today in service of white supremacy. He also urges us to consider the "less direct," that is to say, the systemic ways that America is still killing Emmett Till through mass incarceration, spatial and environmental racism, and the war on drugs. The book thus functions as a social justice text in that it asks readers to consider how the inequalities experienced by black communities today are rooted in the injustices of the past. It is a brutal lesson in history about the de-valuing of black lives—black lives have never mattered, and still don't.

Likewise, albeit indirectly, *Killers of the Flower Moon* invites readers to draw parallels between the long history of exploitation of Native American land and resources described in the book and the injustices still faced by Native Americans today in the pursuit of capitalist profit. On his 2015 visit to Oklahoma, Grann recounts local

Osages' dismay upon finding out that windmill turbines had been erected near Pawhuska. He writes:

> More than a hundred years after oil was discovered in Osage territory, a new revolutionary source of energy was transforming the region. But this time the Osage viewed it as a threat to their underground reservation. "Did you see them? Red Corn said of the turbines, when I returned. "This company came in here and put them up without out permission."
> 299

A subsequent lawsuit was filed by the federal government on behalf of the Osage claiming that the windmills violated the terms of 1906 Allotment Act, but the courts eventually sided with the energy conglomerate financing the project, citing the fact that the company was not mining underground resources owned by the Osage and therefore was not in violation of the act. This story serves to illustrate the fact that Native Americans are still being exploited today in the interests of "big energy" capitalism.

Similarly, though both *Missoula* and *Examen* focus on localized instances of sexual violence, they directly and indirectly invite readers to view these as case studies of a wider culture of rape. In the first chapter of *Missoula*, Krakauer warns readers about the dangers of viewing *Missoula*, which received the unfortunate and misleading title as the "Rape Capital of America," as an outlier in the United States. Krakauer writes that what happened in Missoula is symptomatic of a larger rape culture in American society:

> In fact, 80 rapes over the course of three years appears to be "on par with national averages for college towns of Missoula's size," as Baker mentioned in her piece. According to FBI's latest statistics, there were an average of 26.8 "forcible rapes" reported in American cities the size of Missoula in 2012—which works out to be 80.4 rapes over three years. In other words, the number of sexual assaults in Missoula might sound alarming, but if the FBI figures are accurate, it's actually commonplace. Rape, it turns out, occurs with appalling frequency in the United States.
>
> 10

Likewise, at several points throughout the series, *Examen* invites readers to acknowledge child sexual abuse in the Catholic Church as a widespread and structural problem. As one survivor puts it:

> In Spain, we need a help network of abused people as well as better institutional help and that society opens its eyes and see that this problem is real, that some people suffer, that it's not isolated, it's structural, that it happens to many people statistically, and nothing is being done to stop it.
>
> (S1:E2: 19:35–19:13)

While both *Missoula* and *Examen* both focus on local cases of sexual violence, they continually advise their audiences to consider sexual violence as a widespread and structural problem around the world.

CONCLUSION: LOCATING KAIROS IN THE NEW TRUE CRIME—The timeliness of these books cannot be overstated. Even when historical in content, these books deal with issues that are still relevant in contemporary U.S. society. People of color still face overt and subtle forms of racism on a daily basis. Indigenous peoples of North America are still being exploited and culturally erased in the interests of big oil. Rape culture is alive and well, not only on campuses and churches around the world, but in Hollywood and Washington D.C., too. Indeed, this is why the trend towards social justice advocacy in contemporary true crime is such a welcome development. Given both its chosen subject matter and its widespread appeal in contemporary society, the true crime genre is at a unique vantage point from which to tell stories of the injustices that have been and are still faced by our most marginalized communities. Yet, all too often, true crime has reaffirmed conservative notions of justice, promoted problematic tropes, and neglected people of color and other marginalized communities in the stories they tell.

This all points to the importance of locating what rhetoricians would call the kairos of true crime. Kairos is an ancient Greek rhetorical term that refers to the "timeliness" or "appropriateness" of a rhetorical argument. It asks of a piece of rhetoric: why now? For me, consideration of kairos is a twofold project that aims to determine both a writerly as well as a readerly kairos. In other words, when I say that consumers of true crime should try to locate the kairos of true crime, what I mean is that consumers of true crime should ask themselves, first, why the book they are reading was written when it was written (What was the writer's kairos for writing about this crime?), and, second, why they are reading it right now (What is my kairos as a reading about this crime?). Through such reflection, readers can avoid (or at least critically approach) the true crime that profits from suffering and promotes questionable tropes. In other words, locating the kairos of true crime can help readers create the ethical, moral, and political contexts for consuming true crime.

REFERENCES

Barker, Chris and Jane, Emma. *Cultural Studies: Theory and Practice.* 3rd ed., Los Angeles et al., Sage, 2016.

Bogart, Laura. "Why our true crime obsession is bad for society." The Week, 31 Jan. 2018,https://theweek.com/articles/736073/why-true-crime-obsession-bad-society. Accessed 14 Dec. 2018.

Bolin, Alice. "The Ethical Dilemma of Highbrow True Crime." Vulture, 1 Aug. 2018, https://www.vulture.com/2018/08/true-crime-ethics.html. Accessed 14 Dec. 2018.

Browder, Laura. "True Crime." The Cambridge Companion to American Crime Fiction, edited by Catherine Nickerson, 205–28. Cambridge: Cambridge University Press, 2010.

Couture, Denise. "True Crime Podcast Leads to Arrest of Ex-Rugby Star Over Wife's Disappearance." NPR, 5 Dec. 2018, https://www.npr.org/2018/12/05/673756194/true- crime-podcast-leads-to-arrest-of-ex-rugby-star-over-wifes-disappearance. Accessed 14 Dec. 2018.

Cruz, Lenika. "The New True Crime." The Atlantic, 11 June, 2015, https://www.theatlantic.com/entertainment/archive/2015/06/true-detectives-serial-the-jinx/393575/. Accessed 14 Dec. 2018.

D'Addario, Daniel. "Inside the True-Crime Boom Taking Over Prestige TV." Variety, n.d., https://variety.com/2019/tv/features/hulu-the-act-true-crime-show-1203196058/. Accessed 29 May 2019.

Dredge, Stuart. "Serial podcast breaks iTunes records as it passes 5m downloads and streams." The Guardian, 18 Nov. 2014, https://www.theguardian.com/technology/2014/nov/18/serial-podcast-itunes-apple-downloads-streams. Accessed 14 Dec. 2018.

Grann, David. Killers of the Flower Moon: The Osage Murders and the Birth of the FBI. New York, Vintage Books, 2018.

Houpt, Simon. "Our addiction to true crime has a human cost." The Globe and Mail, 11 Oct. 2018, https://www.theglobeandmail.com/arts/article-true-crime-is-booming-as-entertainment-how-do-victims-and-families/. Accessed 14 Dec. 2018.

Krakauer, Jon. Missoula: Rape and the Justice System in a College Town. New York, Anchor Books, 2015.

Kroll, Justin. "Leonardo DiCaprio and Martin Scorsese to Reteam on 'Killers of the Flower Moon'." Variety, 24 Oct. 2018, https://variety.com/2018/film/news/leonardo-dicaprio-martin-scorsese-killers-of-the-flower-moon-1202992316/. Accessed 14 Dec. 2018.

Lynch, Jason. "Over 19 Million Viewers in the U.S. Watched Making a Murderer in Its First 35 Days." Adweek.com, 11 Feb. 2016, https://www.adweek.com/tv-video/over-8-million-viewers-us-watched-making-murderer-its-first-35-days-169602/. Accessed 29 May 2019.

Miles, Jack. "Imagining Mayhem: Fictional Violence vs. 'True Crime.'" The North American Review, vol. 276, no. 4, 1991, pp. 57–64. JSTOR, JSTOR, www.jstor.org/stable/25125321.

Reynolds, Emily. "Neflix is releasing four brand new true crime documentaries." Shortlist.com, 28 May 2019, https://www.shortlist.com/news/netflix-is-releasing-four-brand-new-true-crime-documentaries-400177. Accessed 29 May 2019.

Rowen, Leslie. "True Crime as a Literature of Advocacy." Undergraduate Theses, 14, 2017, https://scholarworks.bellarmine.edu/ugrad_theses/14.

Storey, John. "What is Popular Culture?" Cultural Theory and Popular Culture. London et al., Pearson/Prentice Hall, 2001.

Tefler, Tori. "True Crime Advice: Is the 'True Crime Boom' a Real Thing? The Most Pressing Questions from the World of True Crime." CrimeReads.com, 27 Feb. 2019, https://crimereads.com/true-crime-advice-is-the-true-crime-boom-a-real-thing/. Accessed 29 May 2019.

Tyson, Timothy B. The Blood of Emmett Till. New York, Simon and Schuster Paperbacks, 2017.

Mentoring in the Middle Years: Major-Based Peer Mentors and an Experienced-Based Sociology Curriculum

Mary E. Virnoche[1*], Alexis F. Grant-Panting[2]

KEYWORDS—*peer mentor(ing), middle years, sophomore slump, high impact practices, student success, retention and graduation rate*

*Corresponding Author: mary.virnoche@humboldt.edu
[1]Professor, Chair
Department of Sociology
Humboldt State University, Arcata, CA 95521
[2]Ph.D. Student, Graduate Teaching Assistant
Department of Sociology and Social Work
Texas Woman's University, Denton, TX 76204

Virnoche, M. E., Grant-Panting, A. 2019. Mentoring in the middle years: major-based peer mentors in sociology. *ideaFest Journal: Interdisciplinary Journal of Creative Works & Research from Humboldt State University* 3: 99–112.

In this article, we present the design and evaluation of a middle years, major-based peer-mentoring program at a small public master's-granting institution in northern California. The middle years encompass the academic period between the end of first year and the beginning of senior year. The peer mentors in this program supported a middle years experience-based curriculum, while offering academic mentoring, career support, and connections to communities and other peers.

We begin by conceptualizing peer mentoring. We then provide an overview of the research literature on high impact practices, as they relate to persistence and graduation. We locate major-based peer mentoring in particular, and middle-years programs more generally, as atypical but effective strategies for supporting student success. We then outline the institutional context and detail a program design that targets key points in a high-impact middle-years curriculum, as well as structures for additional mentor outreach.

The outcomes discussed in this article are based on an online survey, as well as institutional and process evaluation data. Overall most students had met with their major peer mentors and reported positive outcomes related to academic success, major integration, and career planning. Students of color were significantly more likely than white students to have met with their peer mentors. First-generation students were significantly more likely than students whose parents had a college degree to have met with their peer mentors. Additionally, students of color were more likely than white students to report that a major peer mentor helped them attend a department event, a measure of academic integration.

PEER MENTORING AND HIGH IMPACT COLLEGE SUCCESS PRACTICES—In this research, we use the concept "peer mentor(ing)" to refer to individuals, processes, and programmatic structures that constitute a system of communication and support for college students. In general, the scope, duration and approaches of peer mentor programs vary (Collier 2015), but the charge of peer mentors is usually to draw on their own life experiences and training to develop holistic supportive relationships with mentees while offering advice, information, and insights (Burke, Sauerheber, Dye and Hughey 2014; Zellner 2008; Benshoff 1993). The model of peer mentoring described in this research adopts this holistic approach and situates the peer mentor as a bridge to resources, as well as a reassuring source of street knowledge—much of the mentor tool kit comes from their own personal experiences with discovering, interpreting, and navigating the university academic culture and administrative structures.

In our work, we defined "peer mentors" as non-tutoring student support staff; they provided bridges to tutoring, but did not provide those services directly. Peer mentor programs have been a central feature in the world of student affairs. Peer mentors are common in residence life, multicultural centers, and health centers. In the research literature, while the use of the term "tutor" was always linked with direct academic support (Cai, Lewis and Higdon 2015; Sloan, Davila and Malbon 2013), some research also used the term "peer mentor" to refer to this same direct delivery of assistance with academic work (Sherman and Burns 2015; Marcoux, Marken and Yu 2012).

Academic department adoption of non-tutoring peer mentors is less common, in general, and unique in its development in a non-STEM department such as Sociology. In the research literature, academic department-based peer mentor research that focused on non-tutoring models was primarily located in the fields of science, technology, engineering, and math (STEM) with recent histories of addressing diversity and retention (Bowling and Taylor 2015; Gross, Iverson, Willett and Manduca 2015; Tenenbaum, Anderson, Jett and Yourick 2014; Hogan et. al. 2017). To that extent, the research in this article contributes to the development, understanding, and evaluation of the unique challenges and opportunities for non-tutoring peer mentor models within academic programs in general, and social sciences more specifically. Furthermore, this work provides a model for leveraging major-based peer mentors to support high impact and inclusive curricular designs (Kuh 2008) that contribute to measures of student success, such as persistence and graduation.

Differences in college student persistence rates are often theorized in relationship to student integration (Tinto 1993). Academic integration includes meeting with advisors, discussing academic plans, participating in study groups and attending career-related workshops; social integration includes going places with friends from school and involvement in campus clubs. First-generation college students experience lower levels of both social and academic integration compared to students with parents who have a bachelor's degree or even some college experience (Nunez and Cuccaro-Alamin 1998). The National Center for Education Statistics (NCES) reports that students of color are more likely than white students to be first generation (NCES 2009). They face complicated higher education navigation challenges that include daily microaggressions (Sue 2010), as well as outright bigotry, while historically white campuses slowly reinvent structures and policies to align with the needs of multicultural and diverse student bodies (Touchstone 2013).

Older models for student support focused on "assimilation." These models problematized "differences" as deficits in experience and values (Guiffrida 2006; Hurtado and Carter 1997). These deficit models of student success directed primary focus to program designs that "fix" students so that they can be successful in relatively static institutional environments.

More recent student success models define diversity and difference as an asset. They value contributions and strengths that every student can bring to their classrooms and programs (Freire [2005] 1970; Martin Lohfink 2005). Experience-based curriculum (Kuh 2008) and peer relationships (Kuh, Kinzie, Buckley, Bridge, and Hayek 2006; Yorke and Thomas 2003) are among the high impact practices that address inclusive student engagement and success. This work on integration has been linked with theory on upward mobility and cultural capital (Bourdieu [1977] 1984) and social capital linked to peer mentoring in higher education (Moschetti et. al. 2018). Some have designed university-wide mentoring programs focused on developing cultural capital through explicit linkages to academic and career networks (Ortiz and Virnoche 2015; Collier and Morgan 2008; Stanton-Salazar and Dornbusch 1995).

One recent pilot study by National Institute of Health researchers examined a peer mentor training model that "embraces and celebrates the cultural diversity within mentoring relationships" (Byars-Winston, Womack, Butz, McGee, Quinn, Utzerath, Saetermoe, Thomas 2018: 86). Researchers found that participants experienced cultural awareness training as valuable to their work with students from diverse backgrounds. Many peer mentor programs are now incorporating cultural competency training to harness student assets and strengths. In 2012, Michigan State University released a mentor training toolkit that emphasized training mentors to better understand their own cultural backgrounds as assets in developing peer mentor relationships (Bottomley, 2019).

Based on American College Testing (ACT) survey data from 258 four-year colleges and universities, students at the second year and into the latter part of their middle years need particular advising and support (ACT 2010:5). This period, sometimes referred to as the "sophomore slump," represents a second major flight period for reasons different from those for leaving in the first year. Students who were successful in their first year need to, in their middle years, develop connections to academic and career paths, build new relationships that are supportive of that path, and engage in "focused exploration" through service and internship experiences that are much more intentional compared to often free-floating, freshmen-year involvement (Schaller 2005). According to Saveliff (2003), while social ties in the first semester were stronger predictors of Educational Opportunity Program (EOP) student persistence, academic (and career) ties were stronger predictors of EOP student persistence in subsequent years (Saveliff, 2003). EOP is a U.S. Department of Education TRIO program that provides comprehensive support for low-income, first-generation students.

A 2005 survey of 382 public and private four-year institutions found that only a little more than 35% of the institutions had at least one initiative specifically directed at sophomores (Tobolowsky and Cox 2007) and very few of those involved peer mentors. These sophomore programs were designed to address the early slump period focused on career planning (74%), major selection (65%), and/or academic advising (61%). About half the programs (46%) incorporated social events for sophomores. Yet only 15% of the institutions adopted models that employed peer mentors for sophomores. Within the California State University system, California Polytechnic State University (Cal Poly) provided a typical example of a sophomore program most common at the national level—they connected students and faculty mentors through a living community (CalPoly 2010) but did not include peer mentors.

The American Sociological Association (ASA) has taken particular interest in studying the academic success outcomes of students in our own discipline. The roots of Sociology and a recent renewed interest in social action (Dentler 2002; Burawoy 2004) provides rich grounds for designing and studying high impact practices like peer mentoring. Spalter-Roth, Senter, Stone and Wood (2010) suggest that major to career transition work is very important for students in majors like Sociology that are missing the obvious vocational pathways of majors like Nursing or even Social Work. Spalter- Roth, Van Vooren, and Senter (2015) found that Sociology is a particularly strong major for helping first-generation students overcome barriers in social and cultural capital.

During the last 20 years, one of the ASA flagship journals, Teaching Sociology, published several studies that discuss mentoring. Yet as observed within broader mentoring research outside our discipline, we found no evidence of major-based undergraduate peer mentoring parallel to our model. Instead, the research in Teaching Sociology focused on mentoring of graduate student teaching that involved peer support (Jungels, Brown, Stombler, and Yasumoto 2014; Wurgler, VanHeuvelen, Loehr and Grace 2014; Hunt, Mair, and Atkinson 2012), as well as teaching collaboration models that involved faculty mentors (Finch and Fernández 2014; Moss and Blouin 2014). Other work on graduate-level mentoring focused on professional socialization (Keith and Moore 1995; Jones, Davis and Price 2004) and preparing doctoral students to move into full-time faculty positions.

The Teaching Sociology research that focused on undergraduate sociology mentoring addressed the benefits of student-faculty relationships in collaborative research (Shostack, Girouard, Cunningham and Cadge 2010; Howery and Rodriguez 2006; Crawford, Suarez-Balcazar, Reich, Figert, and Nyden 1996). At the undergraduate level, Howery and Rodriguez (2006) noted that collaborative research provided a "fruitful" context for student-faculty mentoring of minoritized students who might not otherwise seek or find mentoring. Other research on undergraduate peer mentoring appeared in the form of "peer facilitators" or instructional leaders for service learning (Chesler, Ford, Galura and Charbeneau 2006) or classroom-based peer learning models (Foster 2015; Petronito 1991). These models rely on the tutor conceptualization of peer mentors that is different than the model in our study. As noted above, we found no other research suggesting the formal adoption of Sociology non-tutoring peer-mentor models like ours.

A SHIFTING INSTITUTIONAL CONTEXT: THE UNIVERSITY SYSTEM, THE CAMPUS, AND THE MAJOR—Humboldt State University (HSU) is one of 23 campuses in the California State University (CSU) system. HSU is the northern-most campus and in a rural setting with 74% of students originating from areas more than 250 miles away (HSUIR 2012). In the fall term of 2012, HSU served 8,116 students through 49 baccalaureate degree majors, 12 graduate programs, and 14 credential programs (HSUIR 2012). In the last several years enrollments have dropped and the campus has experienced a series of deep budget cuts.

We launched the major-based peer mentoring program as the campus engaged with a CSU system-wide retention and graduation initiative (GI 2025). The initiative called for the closure of student opportunity gaps, as well as improvements in retention rates and decreases in years to graduation. During this same period, while first-generation, as well as Pell-eligible students, continued to make up more than half the HSU student population, the campus increased enrollment of minoritized students. In 2013, the campus earned the federal designation of "Hispanic-Serving Institution (HSI)" (U.S Department of Education 2016). By 2018, students of color made up almost 50% of the student body. Throughout this period, the Department of Sociology maintained greater student diversity than the broader campus. With majors in Criminology and Justice Studies (CJS) and Sociology (SOC), by 2018, more than two-thirds of the students in the Department were first-generation college students. More than two-thirds received Pell grants. And more than two-thirds of students in the Department identified as students of color.

HSU students encounter a conflicted campus environment and community. University classroom windows frame vistas of a redwood forest, a local marsh, and the Pacific Ocean. These sites fuel the curiosity and commitment of Department faculty and students engaged in courses like Environmental Crime, Forests and Culture, Social Ecology, and Environmental Inequality and Globalization. At the same time, the rural California community that surrounds the campus is predominantly white. The red terracotta rooftop tiles across campus buildings signal a history marked by colonialism. That history includes the genocide of the Wiyot people on whose land the university now towers. This continued dark side of the community is evidenced in news headlines that report student and community fear and anger with racism and violence, including the 2017 murder of David Josiah Lawson who was a second-year CJS major at the time he was stabbed at an off-campus party.

Students in the CJS and SOC majors experience a department with a long history and strong identity in social justice and change. In that way, courses like Community Organizing provide tools for one way to address the world around them. Faculty and students are engaged in community action research, activism, and applied sociological research. In much of the coursework and in related applied research and activism, race, class, gender, sexuality, and citizenship are organizing frameworks that drive us in posing challenging questions about power and privilege, and help in developing solutions that support social justice. To some extent, the academic classrooms in the Department provide space for students to understand, discuss, and work together on addressing injustices around them.

Like many programs, during the last 20 years the tenure density of the Department has decreased and we rely more and more on critical "temporary" faculty members. In 2001 the department employed nine tenure-line faculty members and few lecturers to serve one relatively small (136) undergraduate major and a master's program. In 2012, we launched a new major in Criminology and Justice Studies (CJS). In 2018, with the additional CJS major and triple the overall Department enrollments, seven tenure-line faculty members and the equivalent of almost five full-time lecturers supported the three programs. In 2013 during the first year of the CJS program, we enrolled 61 CJS majors and in the next academic year, that number more than doubled to 149. Since 2012, the MA program averaged 20 graduate students (HSUIR 2015) and major enrollments grew. At census fall 2018, we enrolled 367 majors across our two BA programs.

PROGRAM DESIGN: PEER MENTORS SUPPORT AN EXPERIENCE-BASED CURRICULUM AND HOLISTIC ADVISING—In January 2013, we launched one of the first "major-based" peer-mentor programs on campus. Alexis Grant, the second author on this paper, was the grassroots catalyst for the initiative. She was already well-networked and respected among her peers in the Sociology community and it was her idea to start a mentoring program. She was a mature transfer student who had experience with training and outreach for AmeriCorps in the Washington, D.C. region. Her work had involved mobilizing young people to be active members in the community, as well as mentoring AmeriCorps members. Her mentoring supported the retention of the corps members in their year-long service commitments.

We designed the Sociology peer mentor program to provide one-to-one mentoring that supported a Department practice of holistic advising, as well as outreach to groups of Sociology majors in key courses. This course outreach focused on pivotal points in the curriculum where we knew peer support could be most beneficial. Peer mentors met with students in a second-year service learning class, a third-year professional development seminar, and in courses where students prepared materials required for enrollment in capstone internship and thesis courses. The curriculum itself was informed by research on high impact teaching and learning, as discussed in the literature review of this article. The curriculum addresses career uncertainties. It also creates connections between developing academic skills, "real life" experiences, and imagined post-academic life.

The mentors also served as social links to support student involvement with major-related activities and academic community building. They encouraged mentees to attend a variety of department events from degree planning workshops to beach bonfires. They also collaborated with the Sociology Student Association on planning events and generating participation. In the following sections, we provide further detail on the curriculum and the role of peer mentors in supporting students in the program.

Service learning second-year course. Since 2010, the Department consistently taught the second year "Social Issues and Action" course with a service-learning component. CJS majors enroll in a parallel "Criminalization/ Inequalities" class. All students work with a faculty service learning and internship coordinator, as well as the university service learning office, to identify a community organization of interest and complete 13–20 hours of service in that placement. Service learning experiences include everything from assisting at the local food bank to playing board games and talking with teens at juvenile hall. As these are short-term community experiences, they usually involve relatively simple tasks while allowing students a small window into the operations of a community group.

Peer mentors work with the faculty coordinator to support Sociology majors as they search for and confirm their service learning placements. Service learning engages students with the community at an early stage, begins to build their network, and structures an opportunity to think about possible career pathways. While some students are already seasoned volunteers, others have never volunteered or are not familiar with the local community. The prospect of cold calling employers can be very intimidating. Therefore, peer mentors were crucial in supporting this process.

Peer mentors introduced themselves to the service-learning students during the first weeks in the semester. They share their own experiences and offer to meet with students one-to-one to brainstorm on service learning opportunities and make contact with organizational staff. They provided students with their contact information and how they can best connect with them. Mentors also used the student sociology club as an avenue for outreach and connection with students. They attended sociology club events, meetings, and outreach events designed for clubs by the university.

Professional development seminar. After several years as a pilot elective course, in 2013 the Department began to require a one-unit professional development seminar for students in their second or third year. We created the seminar in response to student feedback that the résumé and career work we had been doing in the capstone course was too little and too late. We are one of only about half (51%) of U.S. Sociology programs that deliver career-related content formally in our curriculum and one of one-third (32.9%) of programs that require students to take the course (American Sociological Association 2018). Integrating key elements of professional development and job search (Hecht 2016) into a required seminar addresses career-mentoring challenges for large enrollment majors. The seminars also address equity gaps that develop around professional mentoring delivery without the structure that curricular integration offers. Furthermore, with a professional foundation in place via the seminar, one-to-one student-faculty advising and peer mentoring can build on that knowledge.

Major peer mentors are required to have taken or be enrolled in the professional development seminar. With these skills in place, they are able to support their mentees in seeking service learning placements, internships, and jobs. The proseminar topics include building résumés and cover letters, as well as networking logic, organization systems, and other skills from elevator speeches to working a room. The Harvard Business School reported that 65 to 85% of jobs are found through networking (Harvard 2012). Students also learn to plan and manage a job search, from creating a system to keep track of all the moving pieces to business correspondence and interview preparation. We partner with career center staff so that everyone completes at least one mock interview. And as a class they talk with a panel of community professionals who share their experiences of hiring new staff members. The panel discussion is usually followed by a reception where students are encouraged to apply the networking skills they learned in class.

In the early phases, we team-taught the seminar to give faculty members a chance to solidify their own comfort levels with not only teaching, but also implementing these professional practices. The instructors also collaborate with professional staff at the career center for mock interview support, as well as computer lab trainings where students are introduced to the latest online internship and job sites.

For students planning graduate study, we offer a parallel seminar with a slightly different focus. We build CVs and statements of purpose. We plan for building research experiences that make students competitive in their graduate school applications. Students also develop strategies and systems for researching and organizing their graduate program search processes, as well as linking networking strategies with that search. They all participate in mock interviews where their peers watch and participate in providing feedback—they learn quickly to identify strong points and areas that need work in peer interviews. During the final graduate school seminar meeting, the class meets with a panel of experts. Most terms we include panelists who speak to PhD, MA, MSW and law school considerations and admissions.

Capstone thesis or internship. The capstone is the final experience-based course. In the semester before their capstone, students must decide and get instructor approval to enroll. The preparation for enrollment requires considerable work. Peer mentors are trained on how to support students in the planning processes, even if they have not yet completed capstone themselves.

Peer mentors accompany the capstone faculty coordinator for thesis and internship classes when they meet with students in key gateway classes: research methods and theory. In this meeting, the faculty coordinator reviews the options for capstone and explains the details of forms and contracts required to secure permission to enroll in the internship class versus the senior thesis course. Here again, peer mentors offer to set up meetings with students to help them with the process. For internship students, this often involves 2–3 informational interviews with area organizations and then development of a contract for at least 90 hours of work. For example, one recent student contracted to help plan and coordinate an annual fundraiser for Big Brothers Big Sisters. For thesis students, they must secure approval on a research proposal. Because of the department emphasis on social justice and public sociology, sometimes senior thesis students work for or collaborate with campus departments or community organizations. On the other hand, because of relationships with research centers on campus, sometimes our students choose to secure research assistantships for their internship experiences. In addition, peer mentors encourage their mentees to attend the final presentations of thesis presentations and internship poster sessions scheduled in the last week of each term.

Supporting holistic advising. The Department integrated major-based peer mentors into three existing advising mechanisms: (i) pre-registration group advising; (ii) pre-registration one-to-one faculty advising; and (iii) walk-in main office traffic where students get answers to general questions about the department and other resources. In 2017, the Department discontinued group advising and adopted a policy for holistic one-to-one advising; all majors meet with a professional or faculty advisor at least once each term. Advisors then release registration system advising holds allowing students to enroll in classes for the next term. Peer mentors offer "pre-advising" in anticipation of busy registration period schedules. Pre-advising helps students make the best use of their faculty advising time. Peer mentors help students identify questions, as well as prepare and review important advising materials, such as their degree plan and paperwork required to register for some specialized classes (internship and thesis). All students in our majors develop two or four-year online degree plans that link directly to degree audits in an online academic records system.

PROCESS EVALUATION: CHALLENGES AND ADJUSTMENTS—Throughout the early development of the program, Mary (the first author on this paper) collected qualitative process evaluation data from peer mentors and mentees, as well as from staff and faculty members. She noted information shared in peer mentor meetings, as well as through informal conversations in the office and the hallways. In this section, we discuss some of these data, particularly as they related to changes we implemented in the program design. We focus on identifying peer mentors, structuring contact, navigating caseloads, and training.

Identifying peer mentors. The peer mentors themselves reported overall positive experiences in their work supporting students in our majors. We learned that the best peer mentors were outgoing and mature students who also possessed a good amount of "street smarts" about getting things done on the HSU campus. Some skills/knowledge could be provided through training, but major mentors really did need an already developed sense of the campus, the department, and the major. In addition, the same student in the major who made an excellent writing tutor (quiet and steady) was not necessarily the same student who would be really outgoing, best received, and trusted as a peer mentor.

Students found it easy to talk to Alexis and she often worked with students referred to her by friends. "You need to talk to Alexis…" was the word on the street. Alexis had come up through the community college system like more than half of the other students in the Sociology major. She worked hard and had excellent grades. As an African-American transfer student, she related to the struggles of other students, including the racial challenges of studying on a predominantly white campus in a very white region of the state. The Sociology major was more diverse than most other programs on campus, yet students still experienced struggles with microaggressions (Sue 2019) and structural racism (Crenshaw, Luke Harris and Lipsitz 2018). In addition to navigating the racial landscape of campus, majors also struggled with the demands of being student athletes and parents. Outgoing mentors built rapport and trust to help students meet a range of needs.

Structuring contact: avoiding over-advising and reaching those who most need support. In the original program structure, we built two main mechanisms to connect mentees with a peer mentor: the first caseload method mirrored the design of the HSU first-year peer-mentoring program (Ortiz and Virnoche 2015). The second mechanism relied on faculty members connecting students with peer mentors. Relying on caseload lists, Sociology peer mentors pulled majors to them through frequent communications in person and via email. The three mentors led by Alexis sent out initial communications and visited classes. From these communications, and targeted communication at advising time, they generated meetings. Alexis in her playful yet assertive style would see a mentee on campus and remind them that they owed her a meeting. The second method involved working collaboratively with the faculty. Faculty members pushed Sociology majors to mentors with emails such as "Could you check in on Jackie Jones? Just ask her how things are going?"

In the second year, where we initiated the caseload lists, students were coming to Alexis and asking if they could switch mentors. Alexis did some mentoring of students off her list, but her workload became overwhelming. It was challenging to manage the demands, as the students knew Alexis was the mentor with the most experience. At the same time, she was mentoring students who perhaps needed less support. At least a few students reported to mentors that there were too many emails and efforts to get them to meet and that we should cut back on those efforts. Some students who had Educational Opportunity Program (EOP) advisors, faculty advisors, and now a major peer mentor said they were "over advised." Yet Alexis told her mentees who were also EOP students that she could really help with "major" advising. Another peer mentor felt that in general, the pre-advising meetings were the time when he had the most tangible outcomes—that maybe we could cut back and focus mentoring on just a few key things. In general, the mentors said that plugging into class structures was the easiest and most accepted way to have contact with mentees and work from there.

Similar "over-advising" concerns were part of broader campus conversations that sought to identify and target students with the fewest advising resources and avoid adding yet another layer of support to students who were already well-supported. While we know that some students have multiple contacts with staff and faculty members, as well as first-year peer mentors, there are other students whose one-to-one contact with even one faculty members was fleeting at best. The question remains regarding how to get students who need more advising to contact the right people and when to leave students alone. With a relatively small department, Sociology faculty members were able to provide some of that direction as they identified students in their classes who needed mentoring support. As the Department grew, identifying students in need of peer-mentoring support became more difficult.

In 2015, at the suggestion of the office manager, peer mentors began staffing the main office 4–6 hours each week. Office hours in a more remote general resource room had been too lonely. This main office availability not only provided relief to the office professional staff, but also gave peer mentors greater visibility. The new structure generated an even more lively Department office culture: many students stopped by to say "hello" and more students could receive immediate extended support from peer mentors.

Creating and navigating peer mentor caseloads. In the first years of peer mentoring, all Sociology majors who had 25–90 units received an email in the fall introducing the major peer mentors and letting them know that a mentor would be contacting them. We divided the lists of potential mentees between two mentors and at least initially paired students who identified as African American or Latino with mentors who identified similarly. The program only employed 2–3 peer mentors in any given year. After that first year and as our enrollments grew, we abandoned matching based on ethnicity or other identities. Regardless of ethnicity, in that first year many students encouraged their friends to talk to Alexis, who identified as African American and had the most experience. In addition, in that first year, even majors who had more than 90 units asked to meet with Alexis and we added them to the list of students who we considered part of our mentoring population.

At the start of the third year, at the request of past transfer students, mentors began outreach via email to transfer students over the summer. Sociology at that time enrolled 30–40 transfer students every fall and the system required that they have all their lower division coursework complete. As a designated transfer student advisor for years, Mary saw many students in shock as they worked out tightly scheduled degree plans that potentially had them graduating in three or four semesters. They had a short period to transition into a new university setting, connect, and move on.

We started the program with mentor-mentee intensive email and personal contact. This was the model for the university-wide, first-year peer-mentoring program. We scaled this back to a model that structured mentor outreach around two particular times and related objectives: (i) September/February mentors focused on getting students to department social events, career and study abroad fairs, and providing transfer student transition support; and (ii) October/November and March/April focused on pre-advising meetings to help students get ready for required meetings with their faculty advisors.

Peer mentor training. Mentors received training before and during their mentoring experience. Because most academic departments seldom have the capacity to fully train mentor staff, we initially looked outside for training programs organized in student affairs. While a centralized training program for student leaders was not fully developed on our campus, several cohorts of major-based peer mentors participated in segments of trainings that were designed for mentors employed by the campus first-year peer mentor program. This spring training included introductions to student development theory, principles of mentoring, campus demographic and retention information, cultural competency skills, leadership, and campus resources. In August, major-based peer mentors also participated in parts of a week-long, first-year mentor training (Ortiz and Virnoche 2015).

In recent years, because of over-taxation of student affairs staff asked to also support our academic mentors, we moved to more limited in-house training. We tried a couple of different models for training shared across 6–8 academic departments with major-based peer mentoring. In general, the shift created instability in training delivery and further highlighted a significant need for a campus-wide infrastructure to support major-based peer mentors.

ONLINE SURVEY EVALUATION OF MAJOR PEER MENTORING—In February 2014, to generate initial assessment data and inform potential program changes, we administered an online questionnaire to 90 Sociology majors on our mentoring list who had received emails and potentially met with our major peer mentors. Some of these students were part of the first year of mentoring and continued in the second year. The response rate was 43%. More than half the respondents (61%) identified as female, a rate slightly higher than female representation in the major (57%). The sample included 34% (14) white students and 32% (11) Latino students. The other respondents identified as bi/multi-racial (3), African American (1), and American Indian (1). Six (15%) students identified as other or did not respond to the question about ethnicity. While African Americans were underrepresented given their enrollment in Sociology, majors identifying with other race and ethnic groups were appropriately represented in the sample, though small numbers make it statistically impossible to draw conclusions based on single racial/ethnic identifications.

More than two thirds (67%) of the respondents identified as sophomores or juniors. While 48% of the students at the university in 2013–14 identified as first generation, 69% (26) of our respondents indicated that they were first generation college students. About half (19) indicated that they worked and went to school.

Mentoring impacted academic success, major integration, and career planning. More than two-thirds (71%, n = 38) of the respondents indicated that they had met with a major peer mentor (TABLE 1). On all but one measure of academic success, major integration, and career planning, at least two-thirds of respondents who met with major peer mentors reported receiving support in these areas. As noted earlier, major integration and career planning are particularly salient factors that maintain student success in the middle years of their college experience (Saveliff 2002; Schaller 2005).

Fifty percent of respondents indicated that a peer mentor helped them attend a department event. Almost all (92%) of respondents indicated a peer mentor helped them feel welcome in the program and more than three-fourths (79%) indicated that they received help from their mentor in connecting with faculty. Likewise, more than two-thirds (67%) reported mentor help with career planning. Overall, the results indicate that the mentor curriculum and outreach was successful in reaching the program objectives.

First generation and students of color more likely to meet with peer mentors. Generation status and ethnic/racial identity impacted the likelihood that students chose to meet with a peer mentor. First generation students were more likely (80%) than continuing generation students (46%) to meet with a peer mentor (χ^2 (1, N = 36) = 4.3, P < 0.05). Likewise, students of color were more likely (81%) than white students (46%) to meet with a peer mentor (χ^2 (1, N = 29) = 3.91, P < 0.10). These outcomes on the use of peer mentoring are promising given reported retention and graduation gaps for these groups of students. As noted in other research, peer mentors can serve as a vital bridge to achieving intermediary objectives of the middle years (academic integration and career planning) that lead to retention and graduation. There was no statistically significant relationship between gender and respondent reports of meeting with a major peer mentor.

Major peer mentoring impacts major integration for students of color. Of the students who met with major peer mentors, students of color were more likely than white students to report that a mentor helped them attend department events. More than two-thirds (67%) of students of color reported support with event attendance compared to 17% of white students (χ^2 (1, N = 18) = 4.0, P < 0.1). Participating in department-related events was one measure of academic integration included in the online survey. As noted earlier, connection to an academic program of study is particularly vital for students in the middle years. These connections support a sense of belonging in a course of study. Coupled with a post-graduation focal point (career direction), these experiences affirm meaning and worth as students face the day-to-day challenges of being a student. Students of color in a predominantly white campus and community often experience additional barriers to academic integration. Major peer mentoring provided particularly promising results for mitigating one of these challenges for students of color.

Connection, security, and guidance: peer mentors "in my corner". The survey included two open-ended questions directly related to major-based peer mentoring. The first asked: "What have been the best aspects of major-based peer mentoring for you?" More than half the participants (55% n = 21) provided one or two sentence/phrase responses. More than half the comments related to the significance of support coming from a peer. Their relationships with peer mentors created a sense of security. Overall, they noted that peer mentors helped them create connections to other students, faculty, resources, and a check that they were on track. One first generation white female transfer student wrote the longest response that captures the sentiment across comments:

> My mentor was willing to stay connected with me during my first semester. This extra resource enabled me to feel more secure in my new surroundings with the Sociology Department. Additionally, my peer mentor met with me in person to listen to concerns that I had regarding my course and gave me suggestions and options. As a result of this regular personal and email interaction, I was able to stay on track and not give up on myself when things got tough during the semester. My peer mentor was genuinely interested in my well-being and how I was doing in my classes and with my instructors. It meant a lot to me knowing that someone was in my corner rooting for my academic success. I am on a clear path towards my graduation at this time…It is a much-needed and invaluable resource.

Another student, a first-generation Latino (male) wrote:

> The fact that you can talk to another student definitely adds another level of comfort and security when inquiring about classes to take in the following semester.

The second open-ended question asked "What

TABLE 1. Major Peer Mentoring Outcomes by Gender, Ethnicity, and Generation Status (n) percentage

	Overall	Gender			Ethnicity			Generation		
		Women	Men	Δ	White Students	Students of Color	Δ	Continuing	First	Δ
	(38)	(22)	(13)		(13)	(16)		(11)	(25)	
Met with a Major Peer Mentor	71	73	62	-11	46	81	+35*	46	80	+34**
	(24)	(16)	(7)		(6)	(12)		(5)	(19)	
A mentor helped me...										
Academic Success										
Be successful in courses	83	75	100	+25	83	92	+9	100	79	-21
Be on track to graduate	75	69	86	+17	67	92	+25	74	80	+6
Make course & graduation plans	71	69	71	+2	67	92	+25	60	74	+14
Major Integration										
Attend department events	50	38	71	+33	17	67	+50*	60	47	-13
Feel welcome in the program	92	88	100	+12	83	92	+9	100	90	-10
Connect with faculty	79	75	86	+11	67	83	+16	100	74	-26
With career planning resources	67	69	57	-12	50	75	+25	100	58	-42

**** $P < 0.05$ * $P < 0.10$ (Fisher's Exact Test – Small Sample Size Adjustment)**

suggestions do you have for improving major-based peer mentoring?" Almost half of the 24 short responses to this prompt actually reinforced the current program design (e.g. "none"; "nothing I liked my experience"). Two to three comments referred to access to and timing of mentoring outreach, opportunities for interactions, and the degree of intrusiveness our program should pursue. They also suggested more mentors, drop-in office hours, and earlier contact for transfer students (implemented and discussed above). They also wanted more social and networking events. Their comments about intrusiveness indicated differing needs in this area: some asked that students be allowed to opt out of continued contact from mentors and others suggested that everyone should be required to participate in at least one in-person meeting with a mentor.

PEER MENTORING AND POSITIVE GRADUATION OUTCOMES—In this section, we discuss institution-level graduation data related to upper division transfer students. We focus on upper division transfer students because they have been a consistent target for the most focused outreach and interventions of our major peer mentor program. Transfer students by definition are already academic success stories—in most cases they have transferred from community colleges and have done well in their coursework. In addition, they have also been successful in navigating the bureaucratic hurdles of transferring to a new institution of higher education. At the same time, transfer students face a compressed timeline for academic integration and career planning. Most transfer students enter our Department and develop a four-semester plan for graduation with an advisor. For them, the reality of such a rapid trajectory to graduation and the next chapter in their professional life is both exciting and daunting.

Sociology transfer cohorts who had major peer mentors had higher graduation rates than earlier cohorts. The 2012 upper division sociology transfer cohort (n = 23) experienced our first and most intrusive efforts at peer mentoring in their second term. They generated a drastic spike (70%) in two-year graduation rates compared to the two-year rates for 2010 and 2011 cohorts (30% and 48%). After that initial spike, on average 52% of upper division sociology transfer students graduated within two years (cohorts 2013–2016) compared to 31% of all HSU upper division transfer students, and 45% of transfer students in the College of Arts, Humanities, and Social Sciences that is home to our Department

In 2013, the Department began accepting upper division transfer students into a new major in Criminology and Justice Studies (CJS). The institutional data indicate that Department-level positive graduation outcomes carried over to transfer students in the new CJS major. Fifty-seven percent of 2013–2016 CJS cohorts achieved a two-year graduation rate. In the periods discussed above, cohort sizes are relatively small (10–36; mean = 22). In addition, it is impossible at the Department level of analysis to untangle positive outcomes potentially linked to major peer mentors from other simultaneous Department interventions and practices (e.g., elimination of group advising, program early adoption of electronic degree planning tools). Comparison data from all-university and college average outcomes suggest more research is needed using university-level data that control for variances in department advising and major-peer mentor practices. While a controlled study of graduation outcomes was not the focus of this research project, the institutional data presented here suggest that major-based peer mentoring should be considered as a factor in future evaluation work.

DISCUSSION AND RECOMMENDATIONS—Overall students reported that contact with peer mentors positively influenced their experience in the Department of Sociology. Most Sociology majors met with their major peer mentors and, of those, most reported that the mentors had helped them across multiple measures— academic success, major integration, and career planning. Students of color were significantly more likely than white students to have met with their peer mentors. First-generation students were significantly more likely than students whose parents had a college degree to have met with their peer mentors. Additionally, students of color were more likely than white students to report that a major peer mentor helped them attend a department event, a measure of academic integration.

During the 2014–15 academic year, with Alexis graduated and Mary on sabbatical, the Department peer mentor program experienced sustainability challenges. A new Department chair added peer mentor coordination to a long list of other responsibilities amidst a loss of faculty members to leaves and retirements. At the same time Department enrollments across all majors peaked in 2014 at 435; recall that the major peer mentor program was launched when enrollment averaged 136. Yet even in this tumultuous period, graduation outcomes remained strong for upper division transfer students in our majors.

Given that positive outcomes on intermediary measures of academic success, academic integration, and career planning took place when the department was much smaller, follow up survey research is planned for 2019. This research will explore the extent to which positive outcomes have been maintained, as enrollments grow and mentors are asked to support greater numbers of their peers. In addition, as noted earlier, more research using institution-level retention and graduation data is also needed. Ideally, this work would control for variance in department-level advising and adoption of major peer mentoring. This work may strengthen the link between major peer mentoring to graduation outcomes. Finally, we recommend qualitative interviews that center student experience of mentoring, advising, and professional development curriculum outlined in this paper. More detailed narrative accounts would lend to better understanding of the nuances of student experiences and related possibilities for adjusting Departmental structures to best support middle year objectives of academic integration and career planning.

Since fall 2015, with six tenure-line faculty back in place and plans for additional hires, Mary resumed coordination responsibilities for the Department peer mentor program. In addition, she resumed consulting on peer mentor program development with faculty members outside the Department. She also provided modest levels of continued cross-program peer mentor coordination. In 2019, as we send this article to press, there remains interest in maintaining peer-mentor programs across several university departments, but there had been no movement institutionally to resource an infrastructure to support these programs.

On the bright side, the HSU 2018 Strategic Enrollment Management Plan directed the campus to build support for transfer students. In addition, Mary worked with the College of Arts, Humanities, and Social Sciences to develop a proposal for GI 2025 funds to support a college structure for major-based peer mentor coordination. Funding of that proposal is pending.

As noted above, a central infrastructure is particularly important for faculty members, typically Department Chairs who have less knowledge about peer mentor program design and need resources for mentor training and support with the day-to-day coordination of peer mentors. The research presented in this paper provides some evidence to support allocation of resources to these infrastructures. Major-based peer mentoring in the Department of Sociology, likely in combination with high impact curriculum and holistic advising, was positively related to academic integration and career planning, as well as improved graduation outcomes.

ACKNOWLEDGMENTS—We would like to thank the Office of Diversity and Inclusion, the College of Arts, Humanities and Social Sciences and the Department of Sociology at Humboldt State University for funding the peer-mentor program. We also thank all the Sociology peer mentors who embraced their roles and made a difference, as well as Sociology undergraduate Zachary Kihm and graduate student Jovanah Arrington for their library research assistance.

REFERENCES

ACT. 2010. "What Works in Student Retention? Fourth National Study: Public Universities and College Report." Retrieved May 18, 2010 (http://www.act.org/research/policymakers/pdf/droptables/PublicFour-YrColleges.pdf).

American Sociological Association. 2018. ASA Research Snapshot. RS2:1.

Benshoff, James M. 1993. "Students Helping Students: A Collaboration Between Counselor Education and Residence Life." College Student Affairs Journal 13(1): 65–70. Retrieved March 25, 2016 (http://eric.ed.gov/?id=EJ475485).

Bottomley, Lisa. 2019. "Mentoring – Part 4: Cultural Competency in Mentoring." MSU Extension. Retrieved May 18, 2019 (https://www.canr.msu.edu/news/mentoring_part_4_cultural_competency_in_mentoring).

Bourdieu, Pierre. [1977]1984. Distinction. A Social Critique of the Judgment of Taste. New York: Routledge.

Bowling, B., M. Doyle, J. Taylor, and A. Antes. 2015. "Professionalizing the Role of Peer Leaders in STEM." Journal of STEM Education: Innovations and Research 16(2): 30–39. Retrieved March 25, 2016 (http://www.jstem.org/index.php?journal=JSTEM&page=article&op=view&path%5B%5D=1908).

Burawoy, Michael. 2005. "For Public Sociology" (2004 Presidential Address). American Sociological Review 70(1): 4–28.

Burke, Monica, Lacretia Dye, Jill Duba Sauerheber, and Aaron W. Hughey. 2014. "Incorporating Religiosity, Spirituality, and Mindfulness into the Professional Development of Residence Life Staff." Journal of College and University Student Housing 41(1): 206-217. Retrieved March 25, 2016 (http://digitalcommons.wku.edu/csa_fac_pub/68).

Byars-Winston, Angela, Verinica Womack, Amanda Butz, Richard McGee, Sandra Quinn, Emily Utzerath, Carrie Saetermoe, and Stephen Thomas. 2018. "Pilot Study of an Intervention to Increase Cultural Awareness in Research Mentoring: Implications for Diversifying the Scientific Workforce." Journal of Clinical and Translational Science 2(2): 86–94.

Cai, Qijie, Carrie L. Lewis, and Jude Higdon. 2015. "Developing an Early-Alert System to Promote Student Visits to Tutor Center."

Learning Assistance Review 20(1): 61–72. Retrieved March 25, 2016 (http://eric.ed.gov/?redir=http%3a%2f%2fwww.nclca.org%2ftlar.html).

Cal Poly. 2010. "Learning Communities: Sophomore Success Program." Retrieved May 25, 2016 (http://www.housing.calpoly.edu/content/res_life/ssp).

Chesler, Mark A., Kristie A. Ford, Joseph A. Galura, and Jessica M. Charbeneau. 2006. "Peer Facilitators as Border Crossers in Community Service Learning." Teaching Sociology 34(4): 341–356.

Collier, Peter J. and David Morgan. 2008. "'Is that Paper Really Due Today?': Differences in First-Generation and Traditional College Students' Understandings of Faculty Expectations." Higher Education: The International Journal of Higher Education and Educational Planning 55(4): 425–446.

Collier, Peter J. 2015. Developing Effective Student Peer Mentoring Programs: A Practitioner's Guide to Program Design, Delivery, Evaluation, and Training. Sterling, Virginia: Stylus.

Crawford, Isaiah, Yolanda Suarez-Balcazar, Jill Reich, Anne Figert, and Philip Nyden. 1996. "The Use of Research Participation for Mentoring Prospective Minority Graduate Students." Teaching Sociology 24(3): 256–263.

Crenshaw, Kimberle, Luke Charles Harris and George Lipsitz. 2018. The Race Track. New York: The New Press.

Dentler, Robert A. 2002. Practicing Sociology. Westport, Connecticut: Praeger Publishing.

Finch, Jessie, and Celestino Fernández. 2014. "Mentoring Graduate Students in Teaching: The FCCIC Model." Teaching Sociology 42(1): 69–75.

Foster, Drew. 2015. "Private Journals versus Public Blogs: The Impact of Peer Readership on Low-stakes Reflective Writing." Teaching Sociology 43(2): 104.

Freire, Paulo. [1970] 2005. Pedagogy of the Oppressed. New York: Continuum.

Gross, Deborah, Ellen Iverson, Gudrun Willett, and Cathryn Manduca. 2015. "Broadening Access to Science with Support for the Whole Student in a Residential Liberal Arts College Environment." Journal of College Science Teaching 44(4): 99–107. Retrieved March 25, 2016. (http://digital.nsta.org/article/Broadening+Access+to+-Science+With+Support+for+the+Whole+Student+in+a+Residential+Liberal+Arts+College+Environment/19 30753/246167/article.html).

Guiffrida, Douglas A. 2006. "Toward a Cultural Advancement of Tinto's Theory." Review of Higher Education 29(4): 451–472.

Harvard Business School. 2012. "Career Development-Networking." Retrieved January 9, 2012 (http:www.alumni.hbs.edu/careers/networking.html).

Hecht, Jaime. 2016. "Preparing for a 21st Century Job Hunt with a BA in Sociology." Footnotes 44(3). Retrieved May 23, 2016 (http://www.asanet.org/footnotes/marchapril16/job-hunt_0316.html).

Hogan, Rosemarie, Deborah Fox, and Georgina Barratt-See. 2017. "Peer to Peer Mentoring: Outcomes of Third-Year Midwifery Students Mentoring First-Year Students." Women and Birth 30(3): 206–13.

HSUIR. 2012. "Fast Facts Fall Semester 2012." Arcata, California: Green State University Office of Institutional Research and Planning. Retrieved September 21, 2013 (http://www.Green.edu/irp/fast_facts.html).

HSUIR. 2015. "Department Enrollment Dashboard" Arcata, California: Green State University Office of Institutional Research and Planning. Retrieved April 21, 2015 (http://www2.Green.edu/irp/Dashboards/Enrollment_Department.html)

Howery, Carla B. and Havidan Rodriguez. 2006. "Integrating Data Analysis (IDA): Working with Sociology Departments to Address the Quantitative Literacy Gap." Teaching Sociology 34(1): 23–38. Retrieved February 16, 2016 (http://www.jstor.org/stable/20058452).

Hunt, Andrea N., Christine A. Mair, and Maxine P. Atkinson. 2012. "Teaching Community Networks: A Case Study of Informal Social Support and Information Sharing among Sociology Graduate Students." Teaching Sociology 40(3): 198–214. Retrieved February 16, 2016 (http://www.jstor.org/stable/41502753).

Hurtado, Sylvia and Deborah Faye Carter. 1997. "Comparing the Determinants of Persistence for First-Generation and Continuing-Generation Students." Sociology of Education 70(4): 324–45.

Jones, Angela Lewellyn, Shannon N. Davis, and Jammie Price. 2004. "Preparing Future Faculty: A New Approach at North Carolina State University." Teaching Sociology 32(3): 264–75. Retrieved February 16, 2016 (http://tso.sagepub.com/content/32/3/264).

Jungels, Amanda M., Marni A. Brown, Mindy Stombler, and Saori Yasumoto. 2014. "Teaching Associates: Bridging Informal and Formal Mechanisms of Support for Graduate Student Instructors." Teaching Sociology 42(3): 220–230.

Keith, Bruce and Helen A. Moore. 1995. "Training Sociologists: An Assessment of Professional Socialization and the Emergence of Career Aspirations." Teaching Sociology 23(3):199–214.

Kuh, George D. 2008. "High-impact Educational Practices: What They Are, Who Has Access to Them, and Why They Matter." Washington, DC: Association of American Colleges and Universities. Retrieved May 10, 2015 (https://www.aacu.org/leap/hips).

Kuh, George D., Jillian Kinzie, Jennifer A. Buckley, Brian K. Bridges, and John C. Hayek. 2006. "What Matters to Student Success: A Review of the Literature." National Postsecondary Education Cooperative. Retrieved May 20, 2015 (http://nces.ed.gov/npec/pdf/kuh_team_report.pdf).

Marcoux, Sarah, Liv Marken, and Stan Yu. "Establishing Peer Mentor-Led Writing Groups in Large First-Year Courses." Collected Essays On Learning and Teaching 5: 55–64. Retrieved March 25, 2016. (http://files.eric.ed.gov/fulltext/EJ1058800.pdf).

Martin Lohfink, Mandy and Michael B. Paulsen. 2005. "Comparing the Determinants of Persistence for First-Generation and Continuing-Generation Students." Journal of College Student Development 46(4): 409–428.

Moschetti, Roxanne V., Scott W. Plunkett, Rafi Efrat, and Dani Yomtov. 2018. "Peer Mentoring as Social Capital for Latina/o College Students at a Hispanic-Serving Institution." Journal of Hispanic Higher Education 17(4): 375–92.

Moss, Alison R. and David D. Blouin. 2014. "Sociology Graduate Student Teaching A Lesson in Long, Strange Trips." Teaching Sociology 42(1): 2–4. Retrieved February 16, 2016 (http://tso.sagepub.com/content/42/1/2).

NCES. 2009. "Table 2. Demography and Enrollment by Parent Education." National Center for Education Statistics. Datalab Tables Library. Retrieved February 12, 2019 (https://nces.ed.gov/Datalab/TablesLibrary/TableDetails/7889?dataSource=B%26B&subjectId=3&topicId=6&rst=true).

Nunez, Anne-Marie and Stephanie Cuccaro-Alamin. 1998. First-Generation Students: Undergraduates Whose Parents Never Enrolled in Postsecondary Education. U.S. Department of Education. National Center for Education Statistics. NCES 98-082, Washington, DC: National Center for Education Statistics. Retrieved January 15, 2015. (http://nces.ed.gov/pubs98/98082.pdf).

Ortiz, Vynessa and Mary Virnoche. 2015. "High School to College Transition-Focused Program: Retention through an Academic Mentoring Program" Pp. 91–98 in Developing Effective Student Peer

Mentoring Programs: A Practitioner's Guide to Program Design, Delivery, Evaluation, and Training, edited by Peter J. Collier. Sterling, Virginia: Stylus.

Petronito, Gina. 1991. "Fostering Peer Learning in the College Classroom." Teaching Sociology 19(4): 498–501

Saveliff, Dan. 2003. "The Hurdles We Overcome: Factors and Processes Influencing the Social and Academic Integration of Non-traditional College Students." MA thesis, Department of Sociology, Humboldt State University.

Schaller, Molly. 2005. "Wandering and Wondering: Traversing the Uneven Terrain of the Second College Year." About Campus 10(3): 17–24. Retrieved January 15, 2015 (http://eds.a.ebscohost.com/ehost/pdfviewer/pdfviewer?sid=95f40fdc-03e3-4628-8891-27d09e2bda8e%40sessionmgr4004&vid=7&hid=4213).

Sherman, Jacob D. B., and Heather L. Burns. 2015. "`Radically Different Learning': Implementing Sustainability Pedagogy in A University Peer Mentor Program." Teaching in Higher Education 20(3): 231–243.

Shostak, Sara, Jennifer Girouard, David Cunningham, and Wendy Cadge. 2010. "Teaching Graduate and Undergraduate Research Methods A Multipronged Departmental Initiative." Teaching Sociology 38(2): 93–105. Retrieved February 7, 2016 (http://tso.sagepub.com/content/38/2/93).

Sloan, Thomas, Federico Davila, and Eleanor Malbon. 2013. "Student-Facilitators as University Tutors: An Effective Approach to Sustainability Education." Australian Journal of Environmental Education 29(1): 80–96. Retrieved February 10, 2016 (http://dx.doi.org/10.1017/aee.2013.16).

Spalter-Roth, Roberta, Mary S. Senter, Pamela Stone, and Michael Wood. 2010. "ASA's Bachelor's and Beyond Survey: Findings and Their Implications for Students and Departments." Teaching Sociology 38(4): 314–329.

Spalter-Roth, Roberta, Nicole Van Vooren, Mary S. Senter. 2015. "First Generation Sociology Majors Overcome Deficits." Washington, DC: American Sociological Association. Retrieved July 4, 2015 (http://www.asanet.org/documents/research/pdfs/First_Generation.pdf).

Stanton-Salazar, Ricardo D. and Sanford M. Dornbusch. 1995. "Social Capital and the Reproduction of Inequality: Information Networks among Mexican-Origin High School Students." Sociology of Education 68(2): 116–35.

Sue, Derald Wing. 2010. Microaggressions in Everyday Life: Race, Gender, and Sexual Orientation. Hoboken, New Jersey: John Wiley & Sons, Inc.

Tenenbaum, Laura S., Margery K. Anderson, Marti Jett, and Debra L. Yourick. 2014. "An Innovative Near-Peer Mentoring Model for Undergraduate and Secondary Students: STEM Focus." Innovative Higher Education 39(5): 375–385.

Tinto, Vincent. 1993. Leaving College: Rethinking the Causes and Cures of Student Attrition. Chicago: University of Chicago Press.

Tobolowsky, B. F., & Cox, B. E. 2007. Shedding Light on Sophomores: An exploration of the Second College Year (No. 47). National Resource Center for the First-Year Experience and Students in Transition University of Sout.

Touchstone, Claire A. 2013. "Tie-Dyed Realities in a Monochromatic World: Deconstructing the Effects of Racial Microaggressions on Black-White Multiracial University Students." EdD, dissertation, Loyola Marymount University, Los Angeles, CA.

U.S. Department of Education. 2016. "Programs: Developing Hispanic-Serving Institutions Program – Title V." Retrieved March 25, 2016 http://www2.ed.gov/programs/idueshsi/definition.html.

Wurgler, Emily, Jane S. VanHeuvelen, Shawna Rohrman, Annalise Loehr, and Matthew K. Grace. 2014. "The Perceived Benefits of a Preparing Future Faculty Program and Its Effect on Job Satisfaction, Confidence, and Competence." Teaching Sociology 42(1): 50–60. Retrieved February 16, 2016 (http://tso.sagepub.com/content/42/1/50).

Yorke, Mantz & Liz Thomas. 2003. "Improving the Retention of Students from Lower Socio-economic Groups." Journal of Higher Education Policy and Management, 25(1): 63–74, DOI: 10.1080/13600800305737

Zeller, William J., ed. 2008. Residence Life Programs and the New Student Experience (3rd Edition). Produced in partnership with the Association of College and University Housing Officers - International. Retrieved March 25, 2016 (http://files.eric.ed.gov/fulltext/ED558984.pdf).

Made in the USA
Monee, IL
05 November 2026

67948875R00069